Mobile Access Safety

Mobile Access Safety

Beyond BYOD

Dominique Assing
Stéphane Calé

ISTE

WILEY

First published 2013 in Great Britain and the United States by ISTE Ltd and John Wiley & Sons, Inc.

ISTE Ltd
27-37 St George's Road
London SW19 4EU
UK

www.iste.co.uk

John Wiley & Sons, Inc.
111 River Street
Hoboken, NJ 07030
USA

www.wiley.com

Library of Congress Control Number: 2012951550

British Library Cataloguing-in-Publication Data
A CIP record for this book is available from the British Library
ISBN: 978-1-84821-435-4

Table of Contents

Introduction

> "With the Internet, to be competitive in the market is to communicate information to the outside world. It no longer consists of forbidding access to an organization's data; it consists of mastering information exchange".
>
> Jean-Philippe Jouas (president of LUSIF)
> Extract from *01 Informatique,* 4 September 1998

Remote access has helped realize one of mankind's most ancient dreams: "ubiquity", because with these new technologies, employees can now access all their company's resources, anywhere, at any time and from any device (PC, PDA, etc.)

This development of "nomadism" is linked to a number of technological improvements, such as the "democratization" of the cost of laptops and the proliferation of Internet access, which is now available even in the remotest of places. However, it also has its roots in economic factors such as the globalization forcing companies to be more efficient and responsive in order to survive in a highly competitive environment. The gains arising from the implementation of mobility solutions are indeed many and varied:

– increases in the productivity of employees, who can continue to work while on the move (on trains, in airport lounges, in hotel rooms, etc.),with customers (online demonstration of in-house software use, etc.), and from home;

– increases in the flexibility of organizations and ways of working due to, for example, the development of teleworking, whether permanent (distributed call centers allowing telephone operators to work from home) or casual;

– decreases in the emission of CO_2 into the atmosphere thanks to a reduction in the number of journeys (BT have estimated that the development of flexible working at their organization saves 12 million liters of gasoline per year);

– continuation of business operations in the case of major crises such as the destruction of the company's premises (by fire, water damage, etc.), epidemics (H1N1, etc.), strikes (rail employees, roads, airports), or any other event that prevents employees from reaching their place of work or to carry out their duties (power cut);

– decreases in the cost of business operations due to the reduction of costs related to the provision of office space (reduction in office space and therefore of rental costs, maintenance, security, insurance, etc.). BT has estimated savings in operating costs of £6,000 per year, per telecommuting employee;

– reduction in transport costs for employees, who can limit work-related travel because they have access to all necessary resources from any given location;

– reduction in employee turnover, by allowing them to adapt the way they perform their professional duties to their own requirements/personal wishes (e.g., childcare during school holidays, sports competitions, charity events, etc.). This improvement in working conditions can increase employee productivity and reduce the number of work stoppages;

– avoidance of fatigue and stress for employees by limiting the travel to the workplace (telecommuting).

Unfortunately, the exponential growth of remote access has completely called into question companies' security methods that have survived thus far. As in the Middle Ages, this mainly consists of building high perimeter walls to protect against attacks from assailants, and to strictly limit and control incomings and exchanges with the outside. But today, organizations' physical boundaries are becoming more diffuse as the development of telework extends its geographical perimeter as well as the number of entry points. A veritable "Pandora's box" has been opened by the growing use of remote access. Thus, one employee can potentially inadvertently contaminate the entire information system of their company by connecting, for example, from home with their personal computer that has been contaminated by their children while surfing illegal download sites. The evolution of organizations' security policies is therefore vital.

Each security issue is unique, because such issues depend on the organization's intended use for its remote access, as well as on its own specific limitations and constraints (financial, technical, etc.) For this reason, it is not possible for us to provide, as part of this work, a universal "recipe". We will try, however, over the course of the following chapters, to give you some ideas, approaches, principles and techniques that will allow you to understand, on the one hand, the risks involved, and on the other, provide you with the means to build a security solution for your particular case.

Our aim is not to produce an exhaustive description of the various security issues and solutions concerning the mobile elements of companies, because as you will have gathered, such a task would require a much longer book. We have therefore chosen to adopt a didactic approach to make the reader aware of the various threats and protection solutions, by giving a concrete example based on an average user, and the various attacks suffered during a "typical day."

Then, we place these attacks in the broader context of the different families of risk. This allows us to then present the tools capable of countering these attacks or limiting their effects.

Finally, we finish with our average user by explaining the protection solutions that should have been put in place to protect him. As the field of security is not solely related to technical issues, we conclude by making the link between the various recommendations with one of the main methodological approaches in this area (ISO/IEC 27002).

Chapter 1

An Ordinary Day in the Life of Mr. Rowley, or the Dangers of Virtualization and Mobility

"Appearances can be deceiving"

Proverb

1.1. A busy day

The day promised to be busy for Mr. Rowley. Upon awakening that morning, he knew it would be punctuated by unexpected events – as usual – but what they would be he didn't know.

It all started after breakfast, when, after his son had already been surfing the Web, he decided to get on with preparing his annual report by logging onto his company's fileserver from his personal computer. Thanks to the VPN[1] Internet access solution which had been installed by his company, he could work from home as if he were in the office. What a gain in productivity! And it was so simple: all that he had needed to do was simply install a small software client on his PC and configure it appropriately.

1 *Virtual Private Network*: virtual private networks typically exist over a public infrastructure such as the Internet, thanks to an encryption solution that ensures confidentiality of data exchange.

Then, because his plane took off at 9am and he was worried there might be heavy traffic, Mr. Rowley hurried out of the house – and found himself at the airport more than three quarters of an hour before boarding. It was not a waste of time, though, since with his business class ticket, he had access to the VIP lounge. He took advantage of the opportunity to download his latest emails on his laptop, using the free Wi-Fi[2] access to deal with them while he was travelling. These little desks for travelers to use were really useful. You could even leave your PC connected and downloading emails, and go to the café to enjoy a coffee and a pastry.

Two hours later, when he had arrived at his destination, Mr. Rowley had dealt with all his emails, and even slept for a little while.

It really was his lucky day. It took barely ten minutes from the airport by taxi to get to his client's workplace. As it wasn't the done thing to arrive at an appointment an hour early, he decided to wait in a small café at the foot of the building. This café also offered free Internet access via Wi-Fi, so our man took the opportunity to order, on an e-commerce site, a fashion doll that his daughter wanted for her birthday.

The meeting with his client went as hoped, and Mr. Rowley could finally close the deal on the new V91 model, which he had been working on for several weeks.

Back to the airport, and as he was early again, he made the most of the VIP lounge, and got on with some work. He took the opportunity to transfer the full list of contacts on his laptop to his new smartphone via Bluetooth[3].

Finally, back at home, he was able to celebrate signing the contract with his little family, with a bottle of champagne.

Just before going to sleep, wanting to check his emails using his smartphone, Mr. Rowley made the unpleasant discovery of the disappearance of his precious device. It had slipped from his pocket in the taxi that took him home, without him having noticed. This perfect day ended on a negative note; he would have to replace it as soon as possible, and

---

2 *Wireless Fidelity*: Wireless Ethernet local area network technology, standardized by IEEE (802.11a, 802.11b, 802.11g, 802.11n).

3 Wireless communication technology (2.4 Ghz) invented in 1994 by the Ericsson company to facilitate exchange of information between devices over short distances.

transfer his contacts from his PC again: a slight waste of time, but he thought no more of it.

1.2. The ups and downs of the day

Mr. Rowley was happy, because he had finally succeeded in convincing his client to sign the contract that was so important to his business, and which assured more than $50,000 of turnover in the coming months.

But he did not yet know, that on that day:

– his credit card details had been stolen;

– he had infected the corporate network with a worm, which effectively paralyzed the entirety of its information systems for nearly six hours;

– the detailed plans for the launch of the new V91 model had been stolen by a rival company;

– all the contact details for his clients and prospects had been stolen.

1.3. What actually happened?

While nothing in the eyes of Mr. Rowley could distinguish this day from so many others he had experienced in the course of his long business career, invisible and ill-intentioned individuals had made every effort to take advantage of his lack of knowledge of information security.

It all started when his son connected to a Website which had previously been attacked by a hacker. Upon visiting the site, a worm[4] was automatically installed on Mr. Rowley's personal computer via vulnerability in the operating system. The worm then took advantage of the IP tunnel that had been established with the company network to propagate there, significantly disrupting the functioning of the information system.

Then, at the airport, when Mr. Rowley left his PC unattended, an employee of a competitor who had recognized him, piqued by curiosity, decided to glance at his laptop. It was then that he recognized the plan for the launch of the new V91 model. The opportunity to obtain valuable

4 A program that spreads from computer to computer by reproducing (duplicating) each time, using means as diverse as email, instant messaging, P2P networks, etc.

information that might hamper the launch of the new product was too good to miss. All it took was to use a USB key to copy all of the desktop files on Mr. Rowley's computer, in just a few seconds.

As for the Internet access point used in the café, it was not provided for customers by the owner, but had been installed by a hacker who knew that by placing a Wi-Fi router in a busy place, many victims could be snared. Those who believed they were connecting to popular Websites (eBay, Amazon, etc.) were unknowingly automatically redirected to a server maintained by an attacker. Taking advantage of this middleman position (see Section 2.2.4 *Man in the middle*) between the user and the e-commerce site, the attacker profited by collecting confidential information, including payment card details.

When Mr. Rowley synchronized his address book between his smartphone and his laptop, he had to input a matching PIN on both devices. But a hacker had installed a PC with Bluetooth sniffing software in the VIP lounge. He knew that this kind of place necessarily attracted people holding positions of responsibility, and therefore having easily marketable, confidential information. By analyzing the traffic exchanged between the PC and the smartphone, he could obtain some of the information necessary for authentication (IN_RAND[5]) and could determine the rest through a brute force attack (PIN[6], BD_ADDR[7]) (see section 2.3.4. *Cracking encrypted data*). Once the authentication key had thus been obtained, it was not difficult to retrieve the desired information from Mr. Rowley's mobile phone.

The bad luck of losing the smartphone in the taxi was the good luck of the next customer, who discovered it, and was even luckier to discover that no protection was in place to prevent access. Mr. Rowley had disabled his passcode protection, deciding that he was wasting too much time repeatedly typing it in.

Mr. Rowley's misfortunes did not end there. Whoever had got their hands on the smartphone quickly realized the value of his discovery: all the emails, business contacts, meeting notes, tender offers in email attachments, etc. He

---

5 Random number used when creating the key for pairing Bluetooth devices.

6 *Personal Identification Number*: numerical password used on mobile telephones.

7 Unique 48-bit address which identifies a Bluetooth device.

knew enough people who would be very interested to know all this information – the world of business is sometimes very small!

Unfortunately what happened to our fictional character during this "very ordinary day" is only a small example of the many types of attacks experienced every day by companies which use mobility solutions. In Chapter 2, we present in detail the main types of threats you might encounter, so that you can better understand the scope of potential attacks, and the inventiveness of hackers.

Chapter 2

Threats and Attacks

"Even the most improbable risk is possible"

Gérard Mestrallet (president of GDF-Suez)

It is impossible to list all the attacks with which mobile systems could one day be confronted because there are hundreds, with new ones appearing every week. We have therefore chosen to present the most important, to give you an idea of the techniques and methods that could be used by hackers against you, so that you can assess these threats and put in place appropriate protection measures.

To assist in your understanding, we have classed these threats into five broad categories:

– *reconnaissance phase:* a set of methods and techniques allowing a hacker to gather information about the target before launching his attack;

– *identity/authentication attack*: a set of methods and techniques which allow a hacker to steal the identity of a machine, a program or a user, in order to use existing authorizations;

– *confidentiality attack*: a set of methods and techniques which allow a hacker to obtain information which is not freely available;

– *availability attack*: a set of methods and techniques which allow a hacker to impair the performance of – or even completely disrupt – a service provided by the target;

– *software integrity attack*: a set of methods and techniques which allows an hacker to hijack or modify the normal functioning of a piece of software, so that it performs functions which benefit the hacker.

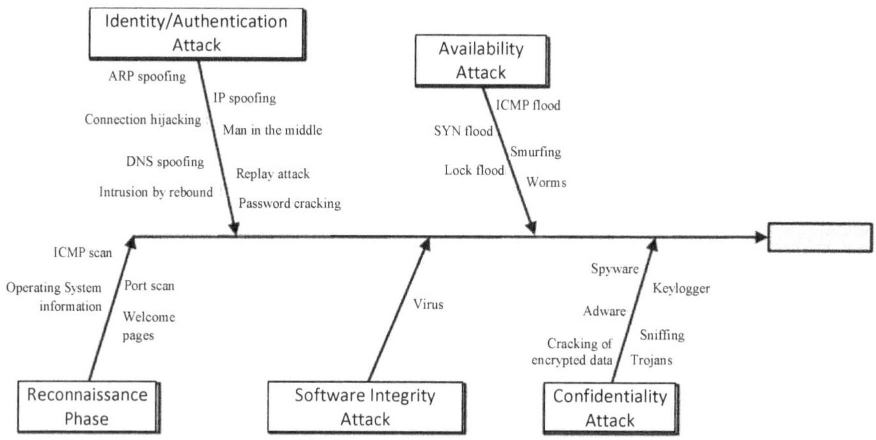

Figure 2.1. *Classification of the different types of attack*

But as you will gather from reading the various paragraphs which return in more detail to these attacks, these five groups are not necessarily set in stone. Indeed, an attack could belong to several of these families. For instance, a "Trojan horse" may obtain information about the system on which it is installed (recognition phase), before stealing the administrator passwords (confidentiality attack) and using them (identity attack) to reformat the hard drive (availability attack).

Equally, you will realize that the types of threats that may affect a laptop also concern smartphones, which increasingly have the disadvantage, due to their small size, of being easily stolen or lost. It is estimated that in France alone, 500 smartphones are "lost" every day. Moreover, in general they have several wireless communication interfaces (Bluetooth, infrared, GPRS, UMTS, etc.), that provide the hacker with many channels of attack.

You will also note that the broad categories of attack we present (except the last, which is specific to new technologies) borrow all of the numerous elements of military strategies that have historically been developed by generals during their campaigns.

In the Chinese strategy treatise, "The Thirty-Six Stratagems", which probably dates from the Ming Dynasty (1368-1644), it is recommended that you conduct a reconnaissance phase with respect to the enemy you intend to attack (knowledge of the terrain, identification of positions, evaluation of strengths, etc.), in order to develop a strategy (打草惊蛇, *beat the grass to startle the snake*). This strategy then consists of causing confusion and disorder in enemy ranks, so that it becomes easy prey (混水摸鱼, *muddy the water to catch the fish*).

Another trick can be used, wherein the attacker takes on the guise of another person in order to cross lines of defense (金蝉脱壳, *the Golden Scarab sheds its shell*). Alternatively, the subterfuge exists not to hide the attacker's identity, but rather to disguise their true intentions as initially innocent, to lull the victim into a false sense of security (笑里藏刀, *conceal a sword in a smile*).

2.1. Reconnaissance phase

In order to prepare an attack, the hacker will use a number of techniques, that will be described further here, to better understand the architecture of the remote access infrastructure and find potential vulnerabilities. In order for mobile devices to be able to access an organization's resources, the organization must provide its users with an infrastructure to process their connection requests. Since, by definition, this infrastructure is located between the outside world (e.g., Internet) and the organization's network, it is a primary target for hackers.

Attackers can collect information via *passive mode* or, equally, *active mode* information-gathering techniques. A combined use of both approaches is common. The use of the term "passive" indicates all technical information gathering where the attacker does not interact directly with the target's information system. In the active mode, direct action on the part of the hacker is required to obtain the desired information.

2.1.1. *Passive mode information gathering techniques*

As in military tactics, the passive reconnaissance phase consists of collecting as much information as possible about the target without being detected. In the Middle Ages, a spy disguised as a simple merchant, positioned in front of the castle, could note all the comings and goings of the garrison, analyze how the fortress was built, and listen to conversations, in order to collect all relevant data.

The approach with respect to a target information system is identical. The great advantage of this passive mode reconnaissance phase is that it is absolutely imperceptible by the target. No special attempt to access systems, and no abnormal or different activity in daily traffic, will be found on any of the target's detection systems.

In the context of a passive mode reconnaissance phase, the attacker can:

1) collect information about the observed subject (an individual, a business, etc.);

2) look for direct vulnerabilities or critical information on the target's information system;

3) listen to network traffic;

4) collect data relating to Web domains.

It should be noted that all the passive mode reconnaissance techniques described here are publicly available, in terms of both tools and services.

2.1.1.1. *Collection of information about the target*

Passive mode information gathering aims to find out all information about the target by querying search engines such as Google, using keywords such as the target's name, email addresses, etc. Very often, valuable data are available in the public domain without the target knowing, or being able to perceive this opportunity from the point of view of the attacker.

Thus, an employee using an Internet forum to post technical questions on issues related to a version of hardware could provide valuable information about the architecture of their information system (potential vulnerabilities, development in progress, etc.).

A simple email address can also be informative about the way in which email addresses are constructed. This then allows fake emails to be forged for use in *phishing* campaigns.

This type of information gathering is often classified as part of the business intelligence domain, but from the attacker's point of view it is primarily to discover all information, clues and weaknesses inside the company, which can then be exploited in order to carry out the attack.

Too often this research phase is dismissed by security services, who forget that apparently innocuous public information can be used to "penetrate the fortress". In this way, by simply conducting an Internet search for the name of Paris Hilton's dog, a hacker was able to recover the entire address book of her smartphone, because the password that protected access to her mobile phone was none other than the name of her favorite pet.

2.1.1.2. *Google Hacking*

Google has established itself as the most popular and most widely used search engine. For most Internet users the use of Google is limited to entering keywords. However, Google has advanced functions that can find information that is sensitive from a security point of view; this is known as *Google Hacking*.

In August 2011, the researcher Tom Parker was able to obtain the SCADA[1] functionality for the administration of certain power plants, accessible on the Internet via Google, by targeting the name of a specific *driver* contained in the headers of HTML pages from Websites referenced by Google.

Although for the layman, the following functions:

– *example 1*:

inurl:"editor/list.asp" | inurl:"database_editor.asp" | inurl:"login.asa" "are set"

or indeed:

– *example 2*: inurl:-cfgintext:"enable password"

1 SCADA (*Supervisory Control And Data Acquisition*: system allowing remote control of technical infrastructure).

are not particularly expressive, they form part of a hacker's basic knowledge. Getting to know their target very often forms a major part of a hacker's capacity to gain access to your systems.

The first example asks Google to search in the URL content for the keywords "editor/list.asp"or "database_editor.asp"or "login.asa" "are set". The symbol | is the equivalent of the logical operator OR.

This search focuses on finding pages of sites where the *login* (username) and *password* are directly accessible in plain text.

The second example conducts a Google search for the keyword "cfg" in URLs and the term *enable password* in the text of indexed pages. This search aims to find the "clear passwords"of network equipment, identified by the "cfg"of URLs.

2.1.1.3. *Listening to traffic network (sniffing)*

Formerly, before the mass deployment of Wi-Fi networks, listening to network traffic required access to the target's premises. The reconnaissance phase was therefore not in passive mode, since it was necessary to obtain physical access in order to connect listening devices. But now, thanks to the rise in Wi-Fi networks, all this can be done from outside your premises, without any need for forced entry. The reconnaissance phase can be accomplished in passive mode from any location offering Wi-Fi access.

Wi-Fi hotspots that offer easy connection to the Internet from a public place often present security risks. Specifically, access through an unprotected public hotspot to messaging facilities that do not incorporate encryption features allow the transmission of identifiers and passwords in plain text.

However, the same problem can occur with Wi-Fi networks which are inadequately protected, and therefore, in the most extreme cases, allow an attacker to record communications and extract sensitive information. And even if the Wi-Fi network has integrated protection mechanisms, a hacker can discover, while remaining perfectly clandestine:

– hardware addresses of devices that pass through this network. On this basis, the attacker can perform a search for the type and manufacturer of the pieces of equipment to check whether vulnerabilities exist and subsequently exploit them;

– the structure of the network. This allows the hacker to manipulate or exploit the addressing scheme;

– the habits of the people who connect (frequency of use, connection times, etc.). The attacker can then use this information to plan the moment when their attack will have the best chance of going unnoticed: the time a given person is away from the network, or conversely, the time when they are present, to avoid raising suspicions.

And even if the network itself is secure, a poorly configured wireless printer is sufficient for all documents sent to it to be captured.

As with military tactical planning, the attacker will collect all possible information to be sorted afterwards, because it is difficult to determine in advance which data will be critical. To assist in this task, there is a broad spectrum of tools, ranging from the most trivial solutions (such as switching the network card to the mode known as *promiscuous* to allow their computer to listen to traffic), to dedicated *open source* software, created by the hacker community.

2.1.1.4. *DNS Analysis*

The DNS (*Domain Name System*) constitutes one of the pillars necessary for the proper functioning of the Internet. To connect clients and servers on the Internet, communication protocols use IP addresses. These addresses, made up of series of numbers, are not the easiest for a person to remember, unlike names. The requirement of establishing an association between these numerical addresses and names has therefore been imposed to allow for easier memorization of sites' addresses. For this reason, we type www.google.com instead of 173.194.34.56.

When you administer a domain name it is in your interest to configure a DNS server, so that you know, for example, how to contact your site or send emails to the administrator. However, in doing so, you provide information to potential attackers. If the DNS server is configured correctly, the information distributed indicates nothing more than how to contact you. But conversely, if poorly configured, poorly monitored, poorly secured, etc., the information stored in the DNS server may reveal:

– the IP addresses of your test servers;

– the totality of your domain information, including internal information, in the case of a configuration error;

– the version of your DNS server. This therefore allows the hacker to establish if this is up-to-date, and if not, to discover opportunities for an attack.

The DNS is a directory, and as for any burglar who targets a given person, the more information available on the target, the more opportunities for burglary. Clearly, it would not cross anyone's mind to enter into a directory "Mr. X lives at Y, entrance extremely overlooked but backyard allowing easy access for entry through the bathroom window"! A poorly configured DNS server provides just such information.

The approaches presented at this stage are only those carried out in *passive mode*, and leave few, if any, traces of information gathering. However, information obtained by these means are not always sufficient for an attack to be prepared, so the attacker moves to a reconnaissance phase known as *active*, in order to confirm his initial findings, or to obtain more data.

2.1.2. *Active mode information gathering techniques*

Active mode reconnaissance is very different, and can be minimally or extremely intrusive upon the target's systems. Several broad categories of active reconnaissance can be distinguished:

– network discovery;

– port scanning;

– homepage analysis;

– vulnerability scans;

– collection of information on the operating system;

– social engineering, too often overlooked;

– analysis of garbage.

The reconnaissance phase therefore begins with OSI layer 3 analysis, to detect the IP addresses of active machines, and continues with a layer 4 analysis to determine which ports (UDP/TCP) are open. We therefore know which applications are available on the identified machines, possibly even if they are protected by a *firewall* or filtering router. The reconnaissance phase then continues with a layer 7 analysis, which attempts to determine which operating systems are being used, and what these machines are for. We finally finish with two reconnaissance methods, which are sometimes overlooked, but are nonetheless relevant: social engineering and analysis of garbage.

2.1.2.1. *Network discovery*

This reconnaissance phase is dedicated to discovering the IP addresses of the target's active machines. Several methods exist, and we present below the most well-known: *Ping sweep* and *Scan arp*.

Ping sweep relies on the ICMP protocol. It consists of sending *echo* requests to all IP addresses of a local subnet. Each active machine, in the absence of a *firewall*-type filter system for example, responds to this request by sending back a return ICMP *echo reply* packet.

This method allows the initial discovery of machines present on a subnet to be achieved quickly and very simply, and also exhaustively, as each possible IP on a subnet will be queried. However, this approach is very "noisy", and with the proliferation of protection such as *firewalls* or antivirus software which integrates alerts for this type of discovery method, the attacker can be speedily identified, and their attempt blocked. Spacing queries over time and the use of a non-sequential ordering of the IP addresses queried are some of the possibilities that the attacker can implement to circumvent protection mechanisms.

NOTE.– In addition to this phase of the investigation *via* ICMP, the attacker can also use the *traceroute* utility, which generates a list of all the equipment that must be passed through in order for two machines to communicate. The structure of a network can therefore be determined, and as a result, the IP addresses of the principal interconnection nodes can be established and used in a Denial of Service attack, for example.

Because *ping sweep* is a "noisy" approach, the attacker may prefer a network discovery technique based on the ARP protocol. ARP scanning consists of sending an ARP request for each IP address in a given subnet. Here we rely on the principles of operation of the OSI layer 2. Each machine uses at least one network adapter to communicate, and each card contains information that uniquely identifies a hardware address. On a given subnet, before beginning to interact via the IP protocol, each machine needs to know the hardware address of its correspondent. The ARP protocol facilitates this dialogue and the correspondence between an IP address and a hardware address.

The ARP scan relies on this mechanism to discover who is present on a specific subnet. This type of scan is carried out much more discreetly, because the tools that implement it never go back up to OSI layer 3 (IP), and do not generate alerts in the majority of protection tools on users' workstations. Furthermore, this type of query is not blocked.

This type of scan works on both Ethernet and Wi-Fi networks; its only limitation is that it can only scan the network on which the rogue machine is connected.

2.1.2.2. *Port scanning*

Port scanning consists of sending requests to a specified range of UDP and/or TCP ports, in order to obtain the list of services (FTP, Rlogin, etc.) that are available on a given machine.

Depending on the outcome of these requests, the state of the port can be deduced. This is achieved based on the analysis of different standards (RFC) and whether or not they are respected in a given vendor's implementation.

Some of the most common methods are provided in Table 2.1.

Protocol	**Type of scan**	**Explanation**	**Result**
TCP	TCP SYN	An SYN packet is sent to each of the TCP ports of the machine we wish to examine, as if to establish a TCP session.	If the return packet is: – a SYN/ACK, the scanned port is open, – an RST, the scanned port is closed, – nothing, the scanned port is filtered.

Table 2.1. *Principal methods of port scanning*

TCP	TCP Connect	The system function connect () is used in order to attempt to establish a TCP session with the target machine on each of the ports in the specified range.	If the target machine responds to this request, the connection process is completed, and then the session is terminated.
TCP	TCP NULL, FIN, XMAS	A TCP packet, without SYN, RST or ACK flags, is sent to each of the ports on the machine we which to examine. These three scanning techniques are only distinguished by their use of FIN, PSH or URG bits. With respect to: - NULL scanning, none of these bits are activated, - FIN scanning, only the FIN bit is activated, - XMAS scanning, the FIN, PSH and URGbits are activated, "lighting up" the packet like a Christmas tree, hence its name.	If the response is: - an RST packet, the scanned port is closed, - an ICMP *destination unreachable* packet, the scanned port and/or the scanned IP address is filtered, - nothing, the scanned port is open or filtered.
UDP	UDP Scan	A UDP packet is sent to each of the port of the machine we wish to examine.	If the response is: - a UDP packet,the scanned port is open, - a*port unreachable* packet, the scanned port is closed, - an ICMP *destination unreachable*packet, the port and/or the corresponding IP address is filtered, - nothing, this indicates that the scanned port is open or filtered.

Table 2.1. (continued) *Principal methods of port scanning*

These basic methods have been the subject of refinements, notably in their covertness, in order to escape detection systems: scans spread over time so as not to attract attention, or generation of IP packets which are not linked to the attacker, to drown the hacker's true IP address in the volume of logs. We conclude with an example that deserves attention. Conceived towards the end of the 1990s (1997, to be precise), this principle has recently seen a new lease of life: the *idle scan*.

In the case of an *idle scan*, the packets which are sent are constructed so that the source IP address is not that of the machine performing the scan, but rather that of another machine (known as a *zombie*). Through indirect observation of the IP packets of the zombie machine and their attached identification number, the attacker can deduce the state of the target's computer's ports. Figure 2.2 depicts the execution of this type of scan:

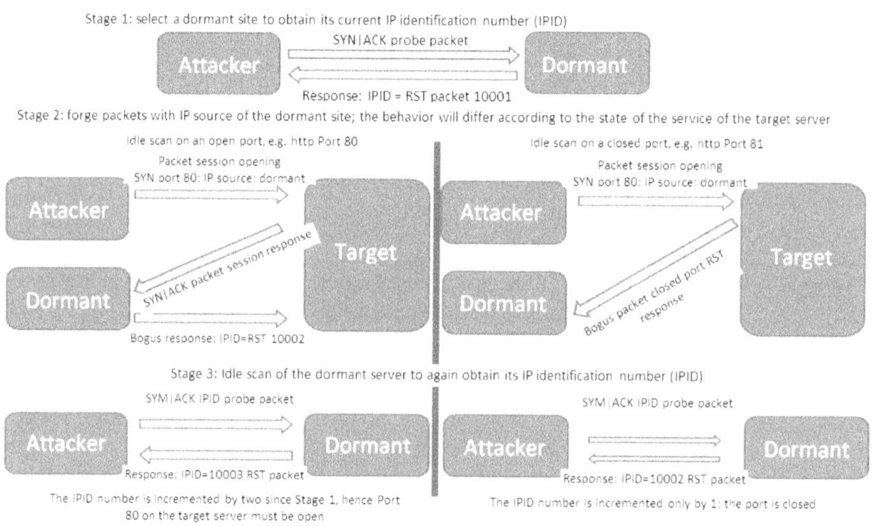

Figure 2.2. *Idle scan process*

Several criticisms of this technique have been made. On the one hand, it is difficult to find a resting zombie machine (the more a machine is active, the more complicated it is to determine the IP sequence numbers). In particular, however, the evolution of different IP layers in operating systems has introduced mechanisms for random generation of these sequence numbers, which complicates the method even further.

Even if the latest OS makes it more difficult to realize this type of scan at the IP layer, this technique has nonetheless found a new field of application in the context of Web 2.0. Because many social network sites allow publication of comments with attached links, this therefore allows these sites and their sharing functions to be used for *idle scanning*, as recently demonstrated by the researcher Martin Obiols.

Thus through social networking sites like Reddit, it has become possible to perform a scan on any port, simply by posting a link, for example to "http://macible.com:8080".

2.1.2.3. *Vulnerability scanning*

Vulnerability scanning differs from port scanning since the attacker not only attempts to find active services (ports), but in particular they will try to determine whether these ports are vulnerable to any existing attack. To do this, a scanner establishes a connection with its target and then sends packets containing the attack code. Depending on the desired setting, the scanner will continue and exploit the discovered fault (in the case of a hacker), attempting to take control of the target, or stop after the simple verification that the application is vulnerable, but not fully take advantage of the flaw detected.

Numerous vulnerability scanners exist, the best known being Nessus, a (partly) *open source* tool. Whether *open source*, commercial or specific to each hacker or group of hackers, these tools methodically identify services and their existing flaws, which they integrate into their internal databases of vulnerabilities. Your security teams can make use of the full range of available tools to test the strength of your defense system against attack and immediate control by a hacker, while remembering that the line between reconnaissance type activities and regular attacks is very thin indeed.

2.1.2.4. *Homepages*

When connecting to certain systems or certain applications, a page or banner appears so that the user can log in (user account, password).

Very often, if the system or application administrator, or indeed the security team, was not vigilant enough, the home page provides valuable information that could be useful to hackers. In such a case, the name of the machine (giving indications of its function) or the version number of the software can be obtained.

Based on such information, the hacker can then, for example, find the list of known bugs related to the software version it has discovered. Thus it becomes clear that all banners must be as neutral as possible, so as not to simplify the attacker's task.

The difficulty of systematic cleaning of this type of information is that it is often a tedious task, sometimes complex (e.g., complete recompilation of the source code of your FTP server and associated non-regression tests), and not always sufficiently documented by all software vendors in terms of how to carry it out.

2.1.2.5. *Gathering information on the operating system*

The operating system is the result of millions of lines of code (nowadays), and the IP stack implementation for each is different. These discrepancies are the result of the way in which each community or vendor has decided to meet, or not meet, or partially meet, the RFCs that define the processing of IP packets. System identification techniques depend on this basis of different IP behavior, and as a result can deduce the type (Windows NT, Linux, etc.) and the version of the OS implemented.

For example, a Windows system which receives a FIN packet returns an RST, while RFC 793 recommends giving no response; its identity is thus discovered. A Linux system responds with a TTL starting at 64, whereas Windows takes the value 128.

It is this collection of differences that provides identification traces for each version and each type of operating system. To automate these analyses, various tools have been created to perform such identifications, and signature databases have been incorporated in the core of these scanning tools. NMAP is the best known, but many others are also available.

2.1.2.6. *Social engineering*

Social engineering consists of abusing a contact by all means possible to obtain key information. In this approach, the attacker uses all human and social vulnerabilities of the contact/target to obtain what they need.

In his book "The Art of Deception", Kevin Mitnick gives examples of this type of approach: based on good faith, naivety or ignorance of employees, some hackers are able to obtain valuable information (passwords, usernames, procedures, etc.). Here, no computer skills are required: all that is needed is a thorough knowledge of human weaknesses along with a good dose of poise and self-confidence.

The attacker first exploits the information they have been able to collect in the preliminary passive gathering phase (obtaining the helpdesk phone number, names of contacts, internal organization of the company, etc.) to gain the confidence of his contact, to convince them that they are a legitimate business person.

The potential of this technique should not be underestimated, especially if at the core of your business no specific procedure has been established to deal with such attacks. In the context of a company whose employees are required to travel frequently (sales representatives, etc.), it is necessary to implement a strict process of personnel identification (e.g., a secret question) when someone asks for a technical service (e.g., a password reset) to prevent an attacker impersonating an employee on the phone, and therefore obtaining the password.

2.1.2.7. *Dumpster diving*

Have you ever been curious about where your company's garbage is stored before collection by the public services? Are you sure that your colleagues use a shredder to destroy any documents containing sensitive information? Do you know if your colleagues in IT, after publishing sensitive architecture designs, destroy them when they no longer need them?

If you answered no to any one of these questions, a quick check can be revealing. Certainly, this is not a prestigious job, but it can quickly become informative: a company's garbage often contains sensitive information (designs, meeting documents, handwritten notes, etc.), due to negligence and/or ignorance. Bins put out on the street are considered *res derelicta*, that is, "things abandoned". It simply remains for hackers, ready to pay personally, to search for and, entirely legally, appropriate them.

2.2. Identity/authentication attack

In order to circumvent security barriers, the attacker also has the option of "disguising" themselves and assuming the identity of a resource that is regarded with total confidence. To do this, there are several techniques that correspond to the principal layers of the OSI model.

2.2.1. *ARP spoofing*

As we saw in section 2.1.2.1, ARP is a communication protocol allowing the resolution of an IP address with the physical (MAC) address associated with each network card. For performance reasons, the majority of operating systems use a cache for storing this information (logical address (IP) <->physical address (MAC)), and therefore avoids having to repeat the resolution request each time an IP packet is sent.

In the ARP standard no authentication mechanism has been put in place to verify the authenticity of the responses obtained; the lack of control at this level allows an attacker to forge, using a technique called ARP *poisoning*, false ARP responses which are sent:

– to computers that can communicate with the attacker's target. This is achieved by overwhelming the IP stack of these machines with false ARP packets: the attacker forces the ARP cache of these machines to be updated with erroneous information. When one of these machines attempts to send a packet to the target, it will therefore also unknowingly use the MAC address of the pirate computer;

or

– even to the attacker's target. In the latter case, the ARP cache of the target machine is forced to update the MAC address of its *gateway* or indeed of its DNS, again by giving the rogue's physical address. In this case, all traffic of the target machine is diverted to the pirate computer.

2.2.2. *IP spoofing*

IP *spoofing* consists of usurping the IP address of a machine that is considered as trusted, in order to appropriate its "identity". Once the hacker has appropriated this identity, he has the same privileges as the victim, allowing him, for example, to circumvent filtering mechanisms (*access list*, *firewall*, etc.), to use the "R" services (Rlogin, etc.) developed by UC Berkeley, or to set up a *man in the middle* attack (see section 2.2.4).

The attacker can also use this technique to commit his misdeeds anonymously, because the logs only record the stolen IP address, and not his true identity.

He can also effect a denial of service, by ending a session. In order to do this, it is sufficient to take the IP address of one of the partners in the communication and send the request to close the session (RST, FIN, etc.) to the other.

There are two different techniques of IP *spoofing*, described in the following two sections.

2.2.2.1. *IP spoofing by modification of the IP address of the pirate machine*

To use the first of these methods, the attacker must ensure that the victim whose identity he wishes to appropriate is not connected or cannot participate in communications. A simple ping command can verify if his victim is connected to the network, and if this is indeed the case, he may, for example, launch a Denial of Service attack (see section 2.4. *Attacks on availability*) so that the target machine becomes unavailable.

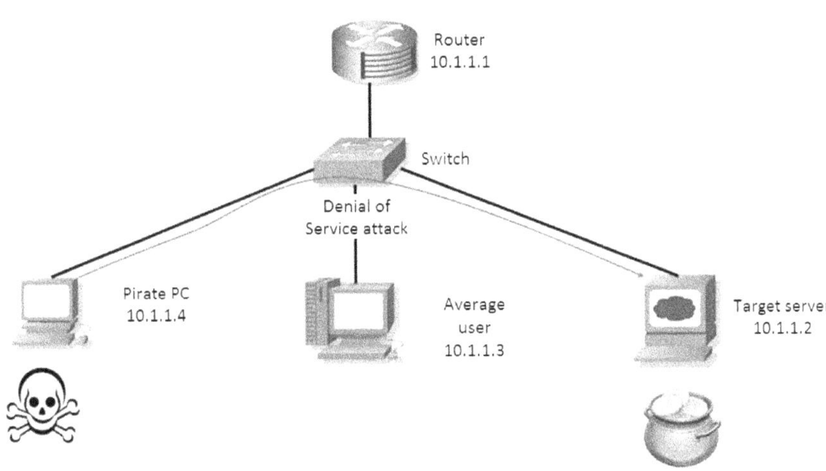

Figure 2.3. *First phase of IP spoofing, denial of service attack on the target machine*

Once the target machine is no longer able to respond to requests, the attacker simply has to change the IP address of his computer.

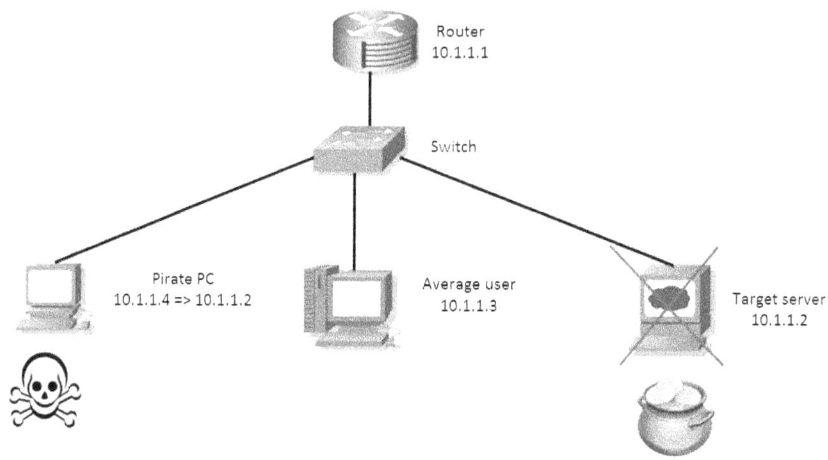

Figure 2.4. *Second phase of IP spoofing, theft of the IP address of the target machine*

Thus, each time a request is sent/destined for the target machine; it is received by the pirate machine. The attacker can then easily access the information sent (passwords, etc.).

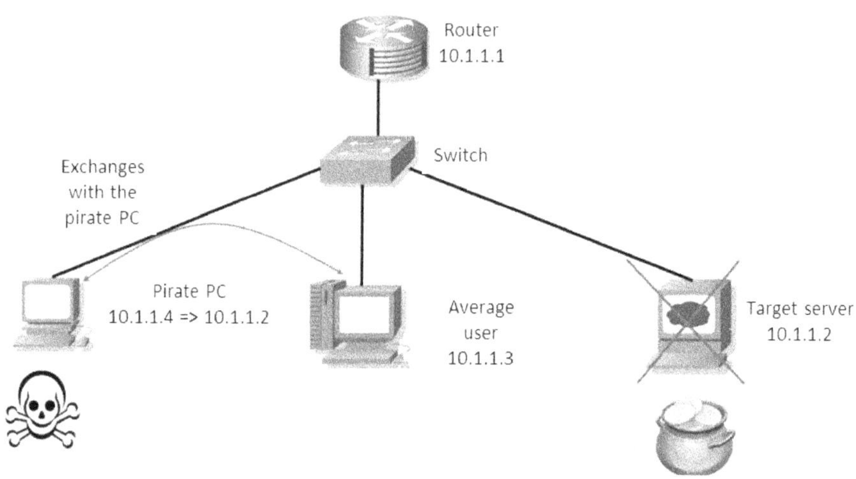

Figure 2.5. *Third phase of IP spoofing, communications destined for the target machine are received by the pirate PC.*

2.2.2.2. *IP spoofing by IP packet forging*

There is another alternative, in which the attacker's machine does not change its IP address, but instead recreates all components of the IP packets with the address that he intends to steal.

2.2.2.2.1. Case where the attacker and the target are on the same network

In the first example of this case, the attacker is in a specific configuration, because he needs to be in a position to monitor traffic that is not directly addressed to him. He may therefore either be connected to a *hub*, or a device that performs the function of a transit node (router, etc.).

Arguably, the use of hubs is not very widespread, which limits the appeal of this method. However, other techniques exist which allow an attacker to turn a *switch* into a hub. For example, when the *Content Addressable Memory*[2] of a Cisco *switch* is saturated, traffic is automatically sent to all ports in the given VLAN when the intended destination cannot be found in the routing table.

Of course, this condition is only true if the attacker intends to obtain information destined for the machine whose identity they have stolen. In the case of a mass mailing of *spam*, for example, they may simply forge packets with the stolen IP address in order to send advertising messages, because receiving responses to these emails will be of no use.

As in the first case, the attacker must ensure that the victim whose identity they wish to appropriate is not connected or cannot participate in communications. Or, if the attacker has succeeded in taking control of a communication node, just needs to simply filter messages to prevent them reaching their intended recipient.

The hacker then simply waits until a message to the target server reaches him, an appeal to which he will respond by building a message from scratch using software that he has previously developed, or more generally by using a *hacker* toolbox.

---

2 Zone in the memory of the *switch* which stores the MAC addresses of Ethernet frames as well as ports of the switch through which they have passed. This allows the *switch* to know on which ports it should forward frames, based on their MAC addresses.

These then automatically input information into the various fields of an IP packet before sending:

– the source IP address (the one which has been stolen);

– the destination IP address;

– source port;

– destination port;

– packet sequence number;

– useful data;

– checksum;

– etc.

Then follow phases 2 and 3, until the hacker has obtained the desired information. In the case of *spoofing* a Telnet server, the attacker need only respond to a connection request by asking for the login and password information in order to receive this precious information in return.

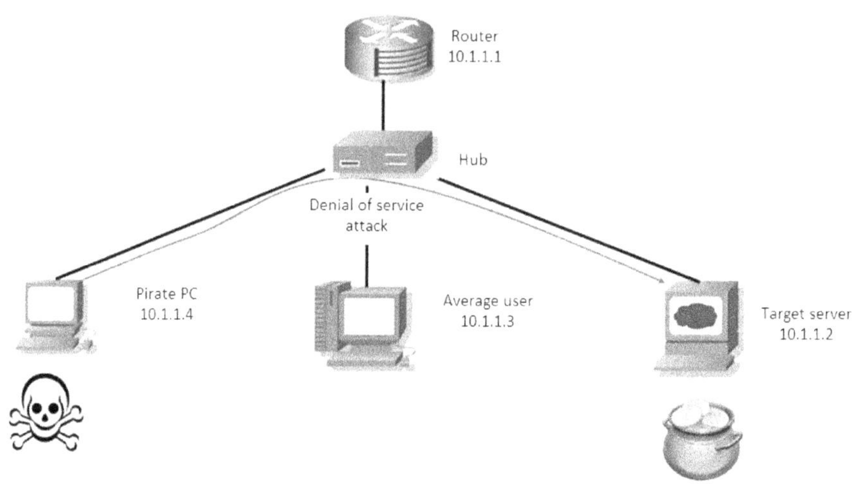

Figure 2.6. *First phase of IP spoofing, denial of service attack on the target machine*

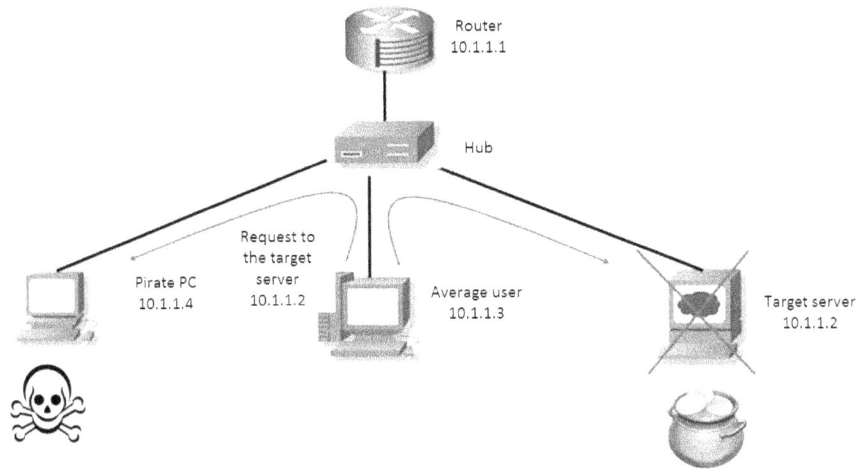

Figure 2.7. *Second phase of IP spoofing, listening to traffic*

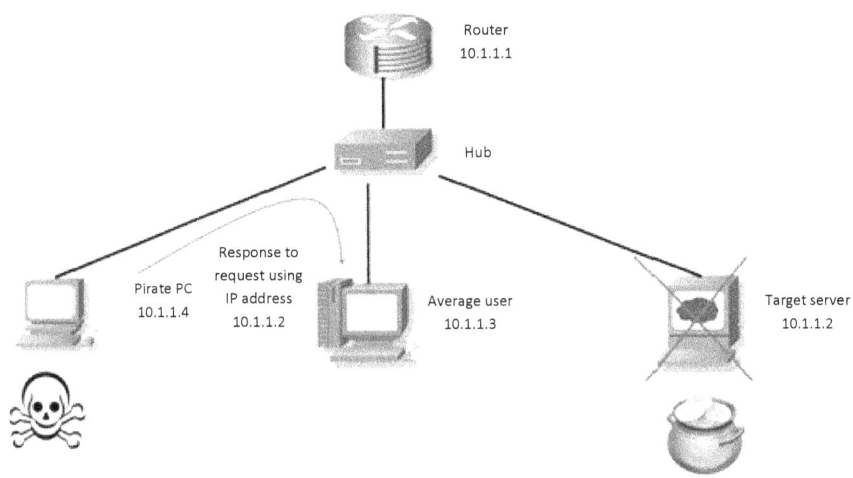

Figure 2.8. *Third of IP spoofing, forging a response packet using the IP address of the target server*

2.2.2.2.2. Case where the attacker and the target are not on the same network

In the majority of cases, the attacker and the target machine are not on the same network. The hacker will therefore not be able to read information destined for the target destination, as it will be sent to the real machine whose identity has been stolen. This is known as a *blind attack*.

Again, before launching an attack, the attacker must ensure that the target machine is "inoperative", because if it receives a response to a request it has not issued, it may, for example in the case of a TCP communication, terminate the session (RST). Once its request is launched, the pirate machine must "predict" the response that the target machine would have made, if the request had actually been received. For example, a TCP uses sequence numbers that must be guessed. In older operating systems, such sequence numbers were determined using only two parameters:

– the time: the counter was incremented by 128,000 every second;

– the number of connections: the counter was incremented by 64,000 for each connection.

This made this exercise easier. However, the system has since evolved, in that most operating systems now use a random number generator to create the sequence number, rendering it unpredictable.

Another option consists of using the *source routing* function, which allows the sender of a packet to specify in its IP header the route by which the answer will be given. The attacker can therefore "force" the receiver to send its data, not to the machine whose identity they have stolen, but to a network controlled by the hacker, where they have put in place a probe (*sniffer*), in order to read the response.

To combat this type of attack, the majority of manufacturers and developers of networks and servers have configured their products to systematically reject this type of query as by default, which greatly limits the scope of this type of attack.

In the same spirit, the attacker may also attempt to modify the routing tables of the network equipment (routers, etc.) situated between him and his victim, so that all responses are transmitted via a path he has specified, rather than the normal route. To do this, the attacker usurps the identity of a routing device and updates the routes, replacing the existing instructions.

To combat this type of attack, routers integrate authentication mechanisms in the dynamic routing protocols (OSPF, BGP, etc.), which prevents this type of table manipulation.

2.2.3. *Connection hijacking*

This attack technique consists of waiting until the victim successfully completes the authentication phase, and then taking his place in the communication. For this to be feasible, several conditions are necessary:

– the victim must be prevented from being able to continue to communicate: a DoS attack is generally used for this purpose;

– they must be able to generate IP packets with sequence numbers compatible with the current session;

– they must target an unencrypted program.

It should be noted that when the attacker launches a DoS attack, the victim may be alerted by the resulting malfunction. Therefore, *hackers* prefer to attack applications in which authentication is based on cookies, because in this case it is possible to be connected at the same time as the victim.

This is the case with the *Firesheep* plugin for the Firefox navigator, which allows an attacker to carry out a *hijack*, notably a Facebook session.

2.2.4. *Man in the middle*

Man in the Middle (MTM) is one of the oldest techniques in data piracy activity. It consists of positioning oneself between the client and the server, and disguising the hacker's machine to the server as a legitimate client, and as the legitimate server to the client itself. By insinuating themselves into the communication in this way, the attacker can then capture all information exchanged between the client and the server, and even change it on-the-fly (the amount of a bank transaction, the recipient of a transfer, etc.). To ensure the success of an attack, the hacker must adapt his behaviour according to the context:

– in the case of a local network, he may use identity theft techniques such as ARP *spoofing* or DNS *spoofing* (discussed in the next section) to change the default gateway or to redirect traffic. A variant of this type of attack involves setting up a malicious Wi-Fi access point and ensuring that victims will connect by playing on the signal strength: a Wi-Fi client always connects to the terminal that offers the best quality service. One of the most wellknown tools for this type of attack is "*dnsiff*";

– in the case of a remote network, the attacker's task is a little more complex: he must first attract his victim via *phishing*. To implement this *Man in the middle* attack, the pirate:

- will need to know the victim's email, as well as the address of the site that he is likely to use;

- will have to build a site that can be mistaken for a real server that the victim has a history of frequenting (online banking site, etc.).

You may have already noticed when checking your emails that MTM attacks via *phishing* campaigns are numerous. Here is an example:

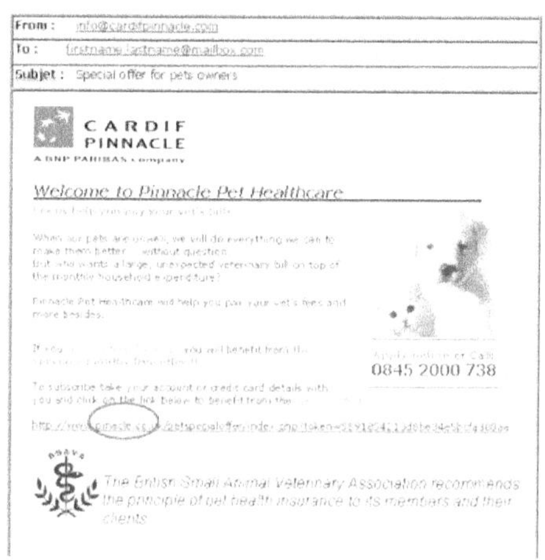

Figure 2.9. *MTM attack targeting the Cardif Pinnacle site*

2.2.5. *DNS spoofing*

DNS spoofing consists of replacing the IP address of a machine referenced in a DNS server with that of a computer controlled by the attacker. When he submits his DNS query, the victim will be automatically – and completely transparently directed to the server controlled by the attacker. A successful attack requires that:

– the server under the attacker's control strongly resembles the site whose identity he "borrowed" (same interface, etc.);

– the site whose identity the attacker intends to "borrow" does not use HTTPS, otherwise an error message during encrypted negotiations could alert the victim to an anomaly;

– the attacker is able to implement the *DNS poisoning* technique, which consists of corrupting the buffer memory of certain versions of domain name servers. These attacks against DNS servers can be made:

- at the level of the DNS root servers of the domain targeted by the hacker;

- at the level of the intermediate DNS servers which cache the responses for certain domains to avoid performance degradation. A variant of this type of attack has recently appeared, thanks to the discovery of a *bug* related to the revocation method of expired domain names in the cache of DNS servers, which allowed hackers to maintain illegitimate (that is, expired) DNS resolutions;

- at the level of the DNS servers used by pre-configured *firewall* packages installed in small offices or mobile workers who connect from home.

2.2.6. *Replay attack*

The replay (or re-issue) attack consists for the hacker of intercepting data packet exchanges during a communication in order be able to subsequently replay certain phases (typically authentication) by re-issuing the apprehended messages in a future session. With some of the first authentication mechanisms, it was enough to intercept the encrypted password in order to later reuse it (without having to decipher it) to gain access to the target system.

The majority of communication protocols now take into account this type of threat, incorporating anti-replay mechanisms that render the majority of such attempted attacks ineffective.

2.2.7. *Rebound intrusion*

When the aim is to attack a system from the outside, the rebound intrusion technique can be used. This serves two purposes:

1) *to avoid* having *to launch a frontal attack on a well-defended system.* A less strategic – and therefore more vulnerable – machine belonging to the same information system as the target is attacked. Or, an attempt is made to go through a lower-security machine that connects to the organization's system via remote access.

Given this compromise, the attacker can branch out along the victim's network, because the security devices put in place are very often weaker when it comes to dealing with communications from areas considered trusted, as the internal network may be.

Democratization of the use of VPN solutions for mobile workstations has reinforced this variant of rebound intrusion; in particular if the configuration of the VPN concentrators allows "split tunneling". In this case, a laptop can open a VPN session with the internal network of the organization, whilst simultaneously having access to the Internet (surfing, etc.), out of reach of all their organization's protection mechanisms. The computer is therefore vulnerable to attack via the Internet, while at the same time being directly connected to the organization's network.

2) *masking the origin of an attack* (source IP address) by carrying out multiple rebounds beforehand via many different machines but also via different countries. This significantly complicates the task of investigators, who must not only find the source of the attack, but also take into account the different laws depending on the countries used by the multiple rebounds.

2.2.8. *Password hacking*

To prove their identity, typically a user must have a password (software, operating system, etc.) which, in most cases, the attacker does not know.

To discover it, they can resort to the "brute force" technique, which takes advantage of the growing computing power of processors and new algorithmic approaches, as we will see in detail in section 2.3.4 – *Cracking encrypted data*. But they may also choose other techniques based on weak passwords, usually chosen by the employees themselves, known as the *dictionary attack*.

When the password choice is left to users, they tend to use easily-remembered sequences of letters and/or numbers, whether:

– names of people they know;

– business terminology (product names, activities, etc.);

– words linked to their private life (first names of their children, husband, wife, pet, etc.);

– common words easily found in a dictionary.

In order to find such passwords, the attacker needs only to compile dictionary databases in the relevant languages, in which the most commonly-used words are stored. The number of passwords to be tested is therefore smaller than in a *brute force* attack and, in many cases, gives good results. In addition, in order to further improve efficiency, the attacker's databases usually include the default passwords installed by manufacturers or software publishers, which unfortunately are not always changed. As an example, in December 2009, the "rockyou.com" Website was the subject of piracy and its user's base of over 32 million passwords was compromised. Analysis of this database reveals which passwords are most popular.

	Passwords	**Occurrence**
1	123456	290
2	12345	731
3	123456789	79 078
4	Password	76 790
		61 958
5	iloveyou	51 622
6	princess	35 231
7	rockyou	22 588
8	1234567	21 726
9	12345678	20 553
10	abc123	17 542
11	Nicole	17 168
12	Daniel	16 409
13	babygirl	16 094
14	monkey	15 294
15	Jessica	15 162
16	Lovely	14 950
17	michael	14 898
18	Ashley	14 329
19	654321	13 984
20	Qwerty	13 856

Table 2.2. *Top 20 passwords used on the "rockyou" Website*

The lesson to be learned from this list speaks for itself! And the analyses carried out in 2011 by Imperva on other password databases gave similar results.

Whatever the approach used to crack a password, if the attack is made via an online system, numerous fruitless login attempts can rapidly be detected in the logs of the authentication systems. In some cases, too many attempts can even block the account, which quickly alerts the user. To avoid being detected, hackers prefer to recover, when possible, the database containing the passwords in an encrypted format, and then try to "crack" them off-line. We return to these techniques in more detail in section 2.3.4 – *Cracking encrypted data*, focusing on the approaches known as *rainbow tables*.

However, mobile equipment too may itself be subject to such attacks. And in the case of smartphones protected by a PIN, this is facilitated by the fact that many manufacturers have integrated a mechanism into the devices (*backdoor*, administrator password, password reset procedure, etc.) that can allow the authentication process to be overridden, so that the technical staff may assist customers who have forgotten their password. Of course, this information has not been slow to be divulgated to the hacker community, which makes these types of protection even more vulnerable.

2.2.9. *The insecurity of SSL/TLS*

SSL (*Secure Sockets Layer*) and *TLS* (*Transport Layer Security*) are encryption protocols which are situated in the session layer in the OSI model, and which offer authentication services, encryption, and key exchanging. In other words, they provide a secure channel between a client and a server.

Historically, SSL was developed by Netscape and used in Web browsers. It underwent two major changes: versions 2 and 3. TLS is the result of the standardization process applied to SSL by IETF, which developed two versions, of which 1.2 is the most recent. In a misuse of terminology, the term SSL is used to designate both SSL and TLS.

To summarize, the main differences between SSL v2 and v3 are as follows:

– the ability to authenticate the client using certificates with V3;

– the use of the compression function required by v3, allowing the structure of the original text to be "scrambled" and making it more difficult to launch cryptographic attacks based on limited knowledge of the structure of the initial plain text messages;

– truncation attacks are no longer possible in version 3. In version 2, an attacker could forge TCP messages to end an SSL session, without the recipient (client or server) knowing that this decision was made by a hacker. The recipient could therefore no longer receive the rest of the messages content, or indeed the attacker could attempt to change the meaning of the final message. With SSL v3 this type of attack is no longer possible. Specific messages at the layer of the SSL v3 protocol must be sent between participants to indicate the closure of a connection. These messages legitimize their sender due to the way in which they are constructed. In fact, the evolution of SSL v2 to v3 is a major step forward in the protection of existing infrastructures.

However, it would be wrong to think that this simple implementation can provide a level of security such that the misuse of information passing through this channel is prevented. As we will demonstrate, attacks on the SSL protocol occur, and allow hackers to gain access to such information. These attacks are based on:

– basic alerts which are ignored by users who either do not understand them or do not pay attention to them;

– the *man in the middle* technique, notably with sslstrip;

– compromising certification authorities, in order to generate false certificates.

The first of these attack strategies requires the hacker to set up a server and pass it off to the victim as the machine which they wish to access. In order to achieve this, as we have seen above, *phishing* or DNS hijacking techniques are available. However, the pirate server cannot present a valid certificate during the SSL handshake, and during the connection an error message (Figure 2.10) will appear on the screen of the victim.

Figure 2.10. *Example of an error message produced during an SSL handshake*

Thus, if the victim ignores this warning and continues the process of establishing the SSL channel, he will find himself connected to the rogue server and will provide them with all the information he wants. In this particular context, no particular action needs to be taken, except to alert the user to pay attention to this type of error message.

The second technique for compromising an SSL connection is based on the *man in the middle* concept. To do this, the attacker must first:

– configure his machine to act as a proxy for all types of traffic, except HTTP and HTTPS;

– implement an *ARP spoof* against the victim's workstation, in order to make believe that the hacker's machine is now the *gateway*. In this way, all of the victim's PC traffic will be directed towards the hacker;

– install software such as ssltrip, in order to redirect all HTTPS traffic to HTTP, giving the victim the illusion that his connection is via SSL, despite this not being the case.

This attack relies on the fact that users rarely enter the URL of a Website by specifying the "HTTPS://" header. Normally, therefore, the connection to the site is first established in HTTP, and later access sections secured with HTTPS. The power of ssltrip lies not in attacking HTTPS, but rather in attacking HTTP traffic. The tool will rewrite all HTTP connections into HTTPS on-the-fly, whilst storing in memory everything that has been changed, in order to maintain the illusion and engender a sense of confidence in the victim. Notably, this includes falsification of "favicon" requests (the image that appears on the left in the browser's address bar) so that it displays a padlock icon, suggesting that it is a secure site (SSL). In reality, unencrypted traffic (HTTP) flows between the victim's workstation and the pirate machine, and the traffic is only encrypted (SSL) between the pirate machine and the legitimate server.

To counter this type of attack:

– first of all, it is important to be cautious and check whether the URL of the site you wish to access has remained in HTTP when it should be in HTTPS. Unlike the first attack technique that we mentioned, in this case there is no error message here to warn you;

– but it is also possible to use plugins in Internet browsers such as "noscript" for Firefox, which forces SSL handshakes for some specific locations.

The third technique for compromising an SSL connection is based on compromising certification authorities.

When an SSL connection is established, the server presents its certificate to prove to the client that it is who it claims to be. The client then carries out a sequence of four checks to ensure the legitimacy of the server certificate presented. It therefore verifies:

– the expiration date of the certificate;

– by which authority the server certificate has been certified;

– the signature of the server certificate, using the public key of the certification authority (available in Internet browsers);

– the name of the domain present in the server certificate, in order to ensure that it is identical to that used by the client to connect to the server.

Attack by compromising certification authorities (CA) consists of creating a false certificate, and to get it signed by a CA:

– by accessing the information system of a CA, to create unbeknown to the client "real fake" certificates, as was the case in 2011, for certification authorities Diginotar and Comodo; or

– by exploiting flaws in a CA's administrative processes (lack of control with respect to the legitimacy of an applicant) to obtain a genuine certificate.

The entire ecosystem of the SSL model is therefore based on mutual trust between the primary authorities whose certificates are present in Web browsers, and the quality of the work of subcontractors, to whom the ability to create certificates is delegated by the primary authorities. In truth, two problems currently exist:

– the exponential growth of the number of certificates from CAs included in browsers (1,482, according to a study by the Electronic Frontier Foundation), which can challenge the legitimacy of all of these certification authorities;

– the quality of work in processing applications for certification is flawed in some cases:

- 6,000 valid certificates have been issued for *hostnames* such as "localhost", according to the EFF;

- certification of the name "Microsoft Corporation" for a person who has nothing to do with Microsoft, and which gave rise to the security bulletin MS01-017.

The case of lawful interception by some governments by creating intermediate certificates should also be mentioned.

All this flies in the face of the trust model on which SSL is based. It is important to be aware that data encrypted in this way is now a commonly-used attack, notably in the case of targeted attacks.

2.3. Confidentiality attack

In our digital society, it is natural that hackers should realize the value that information can represent, and attempt to capture it by developing specific techniques or tools.

2.3.1. *Espionage software*

Among these tools, this section will describe several families of software designed for easy information theft.

2.3.1.1. *Spyware*

Spyware has become a threat with the democratization of the Internet, associated with a significant rise in the exchange of data between computers connected to the network of networks.

These programs are characterized on the one hand by their ability to install themselves without the user's knowledge, and secondly, by having the goal of spying on the computer on which they have installed themselves. Depending on the hacker's eventual aim, *spyware* can collect very specific information (login/password, etc.) or collect everything that is typed on the keyboard of the computer and transmit it to the attacker via the Internet.

Such spyware, originally developed for stations using the Windows operating system, has now extended its scope to the world of smartphones, as they have experienced tremendous growth in recent years, thus becoming a target for the creators of spyware.

With a smartphone, *spyware* may transmit to its creator different types of exchanged messages (SMS, email, etc.), telephone conversations, and even listen to the discussions that took place near the handset using the integrated microphone.

Nothing could be simpler than listening to telephone conversations. The spyware can either:

– send an SMS to the hacker whenever a number is dialed or a call is received, so that he can secretly listen to the conversation;

– record the conversation and send the recording to a server so that it can be listened to later.

Along with worms, *spyware* is one of the online scourges which have developed exponentially during recent years. At the beginning of 2005, a study by *Webroot Corporate SpyAudit* showed that 55% of the selected panel of PCs were contaminated with *spyware*.

The impact of *spyware* in terms of risk to data confidentiality can be extremely high, as demonstrated by the case in 2005, of an employee of a Japanese company who inspected nuclear plants, and whose data (notably, photographs) were found posted on the Winnyfile sharing network, due to a *spyware* infection on his personal computer that he also used for his work.

The development of the use of cellphones has led to software that allows the user to spy on unwitting or unfaithful spouses or unruly children. Once installed on the mobile device, these tools collect information (SMS, emails, call logs, data from integrated GPS systems, photos, etc.) and can even, in some cases, listen to conversations in real time or after the fact, and transmit them discreetly to a jealous spouse or worried parents.

It is also possible, with certain optional operator services, to track the movement of the owner of a mobile phone and know their location at all times.

2.3.1.2. *Keylogger (keystroke recording)*

Keyloggers are a branch of the family of *spyware* due to their shared operating principles. This type of program is designed to secretly record information typed on the keyboard of the computer on which it is installed, and then to surreptitiously transmit it to the attacker. Since their introduction, keyloggers have undergone various technical improvements:

– more intelligent functioning: instead of recording everything that is entered from the keyboard, some keyloggers are triggered under certain conditions, for example, connecting to a particular site, capturing passwords, recognition credit card number formats;

– development of specific hardware keyloggers connected to the cable keyboards;

– recording the victim's session in video format: this circumvents virtual keyboard protection devices which use a mouse to input the password by clicking on icons representing letters and numbers.

Like *spyware*, *keyloggers* also saw a significant expansion with the advent of smartphones. Contrary to what you may at first think, a *keylogger* can be a very simple program written in VB script, and which will not be detected by antivirus software.

2.3.1.3. *Adware*

Adware, in the broadest sense, is any type of program that displays advertising banners while it is operating. The display of various advertising popups is the counterpart to using the software without restriction in functionality: the developer is paid for the advertisements displayed on the user's workstation.

However, this type of software may also collect personal information about the user and transmit it to companies specializing in online marketing. And it is this capacity to collect data, depending on whether this is carried out with or without the user's consent, which leads to the classification of *adware* in the *spyware* category. Some companies "play fair" by clearly informing users about the collection process; Kazaa, for example, specifies in its terms and conditions that its service is free in exchange for the opportunity to collect personal information. However, other companies do not behave this way, either collecting data without the user's explicit consent, or by continuing to collect despite the user's refusal. In this case, *adware* clearly falls into the category of *spyware*.

2.3.2. *Trojans*

The term *trojan* is inspired by Homer's Iliad. It generally refers to software which presents itself in an apparently innocent format (game, utility, etc.) but which hides malevolent functions up its sleeve, and when it is executed can carry out all kinds of tasks without its victim's knowledge: destruction of data, theft of data, use of Internet bandwidth passing as the user for illicit downloads, etc.

A *trojan*'s functionality is limited only by the technical capabilities of the hacker, and by his imagination. Table 2.3 gives an overview of various *trojans*, with their aims and associated functionalities.

Objectives	Functionalities	Characteristics	Example
Extortion (*Ransom ware*)	Encryption of files on the infected machine	Demands a ransom in exchange for the decryption key	Trojan.Pgpcoder

Table 2.3. *Examples of trojans*

Commercial or industrial spying	*Keylogger* and remote control	In May 2005, the Tel Aviv police proceeded to arrest executives of several companies in sectors as diverse as telecommunications, mineral water and the automotive industry, who had used a spyware in order to steal competitors' confidential information	Operation *Horse Race* in May 2005
Embezzlement	Taking control of a mobile telephone	Sending an SMS to surcharging numbers, from which the call cost is deposited into the hacker's bank account	
Remote control	Remote control of the infected computer	Total control from anywhere in the world, for all kinds of transactions (transfer of data on the hard disk, participation in a Denial of Service, etc.).	1998 *Back Orifice* developed by the CdC (*Cult of the dead Cow*[3]) group
Sending spam	Exploitation of Websites offering free sending of message transmission via infected machines	Sending of unsolicited publicity SMS	Delf-HA
Defense of certain values	Adverts based on a user's *surfing* activities	Showing extracts from the Qur'an when a user surfs sites of a pornographic nature	Yusufali-A in 2005 in Korea
Cyber warfare	Reprogramming of the SCADA infrastructure Theft of SCADA information	Development assumed to be at state level. Seemed to target Iranian nuclear facilities. Classified as a worm, but given its ability to receive instructions, can also be viewed as a trojan. Similar to Stuxnet. Again classified as a worm,but able to receive instruction, so may therefore also be considered as a trojan.	Stuxnet[4]June 2010 Duqu September 2011

Table 2.3. (continued) *Examples of trojans*

3 The name "Cult of the dead cow" was chosen with reference to the speech given by the prime minister of China during his meeting with Bill Gates.

4 Stuxnet was developed jointly by US and Israeli services, according to an article in the New York Times (June 2012).

NOTE.– It is important to be vigilant about the use of *freeware* and *shareware* which can easily be found on the Internet, as their mode of distribution make it easy to conceal a trojan, without fear of being discovered. For example, a few years ago the program PKZIP300.ZIP was not as one might think a new version of the popular data compression utility, but a trojan that erased data from the hard disk on which it was installed.

2.3.3. *Sniffing*

It is during transport between the mobile device and the remote access infrastructure of your company that data are most vulnerable, and because the Internet is, by definition, a public area, anyone can try to appropriate or corrupt them.

Sniffing consists of using a probe (which can be hardware or software) to listen to and/or capture network traffic in order to extract confidential information from it (passwords, account, etc.). This technique has long been one of the most used by hackers, since many applications originally did not taken into account security requirements. For example, administration tools like Telnet and Rlogin do not include encryption of identifiers or their associated passwords in their design.

The popularization of the use of Wi-Fi technology has increased this problem, because all data transmitted on the radio waves can be easily intercepted. This is also the case for WPANs (Wireless Personal Area Network), which allow mobile devices to exchange many types of information (address book, etc.), through infrared or a Bluetooth connection, and have the disadvantage of providing only rudimentary protection.

In addition, it is interesting to note the existence of a marketplace selling autonomous, lightweight, low-cost (less than $50), compact equipment which incorporates a comprehensive set of tools to sniff and crack Wi-Fi networks. Designed to be placed surreptitiously near or in the target's premises, these facilities behave like small, passive "drones" for listening to Wi-Fi networks.

2.3.4. *Cracking encrypted data*

With respect to an encrypted message, two main approaches exist for attempting to read it:

– the first, based on a formal approach, attempts to find the algorithm used to produce the encoded text, and then searches for a fault in it, to extract data;

– the second, based on a very practical approach, but not excluding highly theoretical work, as we will see later in this chapter, is to decipher the text by trying all possible combinations of keys: this is the *brute force* technique.

This second approach is discussed here in more detail.

To be truly effective, the brute force attack requires great computing power in order to be able to test all possible combinations in a reasonable time for the attacker.

In recent years, the computational power offered by computers has increased steadily, and helped break certain passwords generated by encryption algorithms previously believed to be inviolable.

Table 2.4 illustrates this fact: it is a list of different code hacking competitions and the results obtained from these challenges.

Organizer	Date	Algorithm used	Result
Cyberpunks Association	1995	SSL with a 40-bit key	An international team (English, Swedish and Australian) and an INRIA PhD student took a total of eight days and a hundred machines running in parallel to break the code.
RSA Data Security	1997	RC5 with a 40-bit key	A Berkeley student succeeded, using 250 workstations in parallel, to decode the message "this is why you should use a longer key" in 3 hours 30 minutes.

Table 2.4. *Results of various "codebreaking" competitions*

RSA Data Security	1998	DES with a 56-bit key	EFF succeeded in finding the message "It's time for those 128, 192 and 256 bit keys" in 56 hours by using a computer specifically built for the task (DES Cracker), which cost $250,000.
RSA Data Security	1999	DES with a 56-bit key	The organization Distributed.net succeeded in finding the message "See you in Rome (2nd AES conference, 22-23 March, 1999)" in 22 hours and 15 minutes, using almost 100,000 computers (including DES Cracker) linked via the Internet.

Table 2.4. (continued) *Results of various "codebreaking" competitions*

Recent years have seen the emergence of approaches and materials allowing brute force techniques to significantly increase their ability to crack passwords. This includes on the one hand, the *rainbow tables* approach, and on the other, the use of the computing power of graphics cards, as well as the capacity of cloud-based processing.

The *rainbow tables* approach is based on a particular principle of pre-calculation of passwords through a *hash* function. This particular structure is derived from the work of P. Oechslin, published in 2003. It is quite wrong to consider a rainbow table as a simple database storing all possible password *hashes*. To understand how a rainbow table works, three points must be considered.

1) *Hash and deduction functions*

A hash function is a mathematical function that transforms a particular entry into a condensed version with the specific property of not being reversible (mathematical proof) or being difficult to reverse (proof by calculation); in other words, finding the input data from the condensed version is computationally impossible given the current state of the art:

– for example, the password "mdpfaible" will have as a hash, with the MD5 function, the following result: 7bf1d0838f4162d4942b2b4130a63488;

– a reduction function, under the *rainbow tables* framework, is similarly a mathematical function that takes the hash of a function as input, and reduces it into a different condensed version;

– a simple reduction function can be to retain just the six initial characters of the previous calculation, that is: 7bf1d0.

2) *The concept of chains*

– at the core of the rainbow table, based on hash and reduction functions, chains are established. A chain consists of the operation illustrated in Figure 2.11. From plain text (starting point), we calculate the hash (operation number 1) which is reduced (operation number 2) and a further hash (operation number 3) which is again reduced (operation number 4) to then be a "hasher" (operation number 5) to give the final result (endpoint).

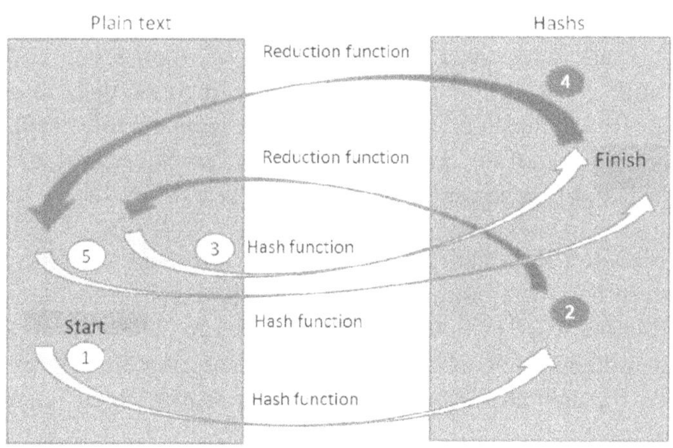

Figure 2.11. *Illustration of the construction of a chain*

A rainbow table will store only the starting point and end point, and the number of reductions to be carried out depends only on the user's preference. This structure can therefore store millions of hashes, retaining only the starting point and destination, and way in which to recover the chain.

3) *The password search algorithm*, based on this particular structure, consists of:

– taking the hash of the password (this is often the only information the hacker uses) and calculating its reduction, then calculating its hash;

– check whether the calculation coincides with an output in the rainbow table;

– if this is not the case, rerun the operations mentioned previously;

– when a result does coincide and the output is found, the rainbow table gives the input; using this, the plain text password can be found.

Rainbow tables are therefore far from being simple databases storing all the hashes of all possible passwords; this would require far too much storage space. This approach is a compromise between reasonable storage space (to retain the inputs and outputs) and acceptable computation time for finding the plain text password on the basis of the construction of chains.

In practice, the calculation based on *rainbow tables* are carried out using powerful computers, and it is then possible to load the results on to a much less powerful machine, such as the hacker's laptop, for example.

The rainbow tables approach has been perfected in recent years to improve its effectiveness and performance, notably via what are known as *probabilistic rainbow tables* based on probabilistic Markov chains.

In another significant advance in attacks attempting to break passwords by *brute force,* is the use of the computing capability of graphics cards and the cloud.

In recent years, the changing needs of image rendering have led graphics cards to offer increasingly significant massively parallel computing capabilities. The idea of using these cards to perform generic calculations appeared recently, and led to the development of what is now known as GPGPU (*General Purpose processing on Graphics Processing Unit*). The provision of language to program the cards for this purpose has also facilitated the development of this approach (including the CUDA *framework* provided by Nvidia, or indeed OpenCL).

Cracking a password by brute force consists of testing many combinations of numbers and letters one-by-one, and by applying a hash function to check for a match with the encrypted password that was intercepted. This type of operation lends itself particularly to the parallel computing framework: carrying out the same type of operation, but with different data.

On this basis, various tools have been developed to exploit these processing capabilities, such as "ighasgpu", an *open source* tool which uses GPGPU to break a diverse range of password types, obviously including Windows NTLM.

Various studies have been conducted over the past two years to verify the effectiveness of this technique, for example those conducted by KPMG[5] or that of Vijay Devakumar. Analyses of the effectiveness of this tool speak for themselves:

Length of password[6]	Time to break
5	1s
6	4s
7	17 m 30 s
8	18 h 30 m
9	48 days

Table 2.5. *Time necessary to break a password based on its length (Source: http://mytechencounters.wordpress.com/2011/04/03/gpu-password-cracking-crack-a-windows-password-using-a-graphic-card)*

In addition it should be remembered that the GPGPU approach can, today, be augmented using the *cloud*, which allows several machines to be combined to work together.

5 *GPU-based password cracking*, Markus Bakker, Roel van der Jagt, University of Amsterdam under the supervision of KPMG, February 2010.

6 Alphanumeric password, with capital and small letters but without special characters.

2.4. Availability attack

A *Denial of Service* attack aims to render unavailable or, compromise the quality of, a service or server. The primary aim is therefore not an intrusion into the target system, but the use of communication channels already open with it, to degrade the quality of service (increased response time).

The technique of this attack consists of saturating the target:

– with requests at the network or transport layer (ICMP, TCP SYN, etc.) that consume all the resources of the TCP/IP stack of the target, and prevent it from responding to legitimate requests;

– with requests at the application layer that take advantage of existing bugs in the target application. A typical example is the flaw that has affected certain versions of Apache servers, (CVE-2011-3192), where with a limited number of HTTP requests, but by completing a particular field, memory and CPU consumption of the Web server increases inexorably, leading to a situation where the service is no longer usable.

From a procedural viewpoint, a denial of service attack can be carried out from a single computer. This is rarely that of the attacker, because he prefers to use a PC controlled remotely. But more generally, denial of service attacks rely on a variant called DDoS (*Distributed Denial Of Service*). This technique involves using a huge number of computers[7], often distributed worldwide, to generate thousands or millions of connections to a particular target. The machines which take part in these attacks are generally computers that have been previously infected with a trojan or a worm, allowing hackers to control them remotely, without legitimate users suspecting anything. A set of machines controlled by a hacker is called a "BotNet", and each PC component is called a "zombie".

NOTE.– If you would like more information about the proliferation of botnets, real-time information on the proliferation of networks of zombie computers is available on the CipherTrust Website.

In 2011 and 2012, there was much media discussion on denial of service attacks, concerning interventions by groups such as *Anonymous* or *Lutezc*,

7 In 2005, the Dutch police arrested a hacker at the centre of a network comprising almost 100,000 zombie machines.

against sites such as Sony or the RIAA (*Recording Industry Association of America*).

Although it is possible to use all kinds of communication channels to perform denial of service attacks, at present, hackers mainly prefer the methods that we now describe.

2.4.1. *ICMP Flood*

This is one of the oldest methods of denial of service, and is the simplest of all. It requires no special technical skill and involves sending multiple ICMP *echo requests* (PING) to saturate the resources of the target. The only requirement for a hacker to successfully complete this type of denial of service is to have a large bandwidth at his disposal.

In order to reduce the nuisance effects of this type of attack, the implementation of *firewalls* that prohibit this type of connection should be considered. In addition, *firewalls* can be configured not to immediately prohibit this type of traffic (ICMP), but to later refuse them if a DoS behavior is detected (e.g., more than a thousand ICMP packets per second).

It is therefore evident that although useful in the context of network systems administration, a configuration prohibiting this type of protocol should be put in place on security equipment, so that it does not itself become the target of such an attack.

2.4.2. *SYN Flood*

Like the attack by ICMP flood, SYN flood is also a very old method. One of its first uses was in 1996 against the New York ISP PANIX, which was unable to provide Internet access for its customers for several days.

The feasibility of this type of attack relies on a weakness of the communication TCP/IP protocol at the transport layer. The normal mechanism for establishing of a connection takes place in three stages:

– first step: the client makes a request by sending a packet in which the SYN field contains a value of 1. This is the request for a new connection;

– second step: the server signifies its acceptance by returning an acknowledgment via a packet in which both the SYN and ACK fields have a value of 1;

– third step: the client confirms the acceptance of the connection by sending a packet in which the ACK field contains a value of 1.

This exchange is known as the "TCP handshake".

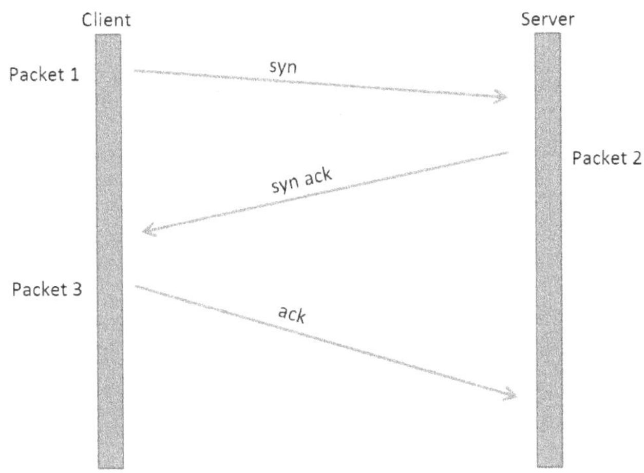

Figure 2.12. *Mechanism of establishing a TCP connection*

However because, for each SYN packet received, the server must temporarily (one to two minutes) keep in its memory the context of the connection request when the SYN ACK packet (acknowledgment) was sent to the client, saturation of resources can quickly occur if a large number of SYN requests arrive in a very short time.

The SYN Flood attack therefore consists of the hacker generating a multitude of SYN packets very quickly and not acknowledging the server's SYN ACK packets, which can easily be achieved on their own machine by configuring their own *firewall* to block any response sent to them by the server.

The hacker thus generates multiple connection requests without completing any of them; the server will then be in a situation where resources are saturated and will therefore refuse all new, legitimate connection requests.

Since the first of these attacks, the SYN Flood technique has improved. Whereas in the past a hacker reduced the possibility of identification of the attack's source machine by falsifying the source address, today he can rely on "Botnet" networks on which the SYN Flood program was installed with or without the knowledge of its owner. Low Orbit Ion Cannon, an *open source* program for performing DDoS, is based on voluntary participants in a botnet, in the context of actions carried out by the group Anonymous.

To counter this type of attack, concerted actions at the ISP level are the most effective. Nonetheless, they remain complex to implement, because they usually require a prior procedure to deal with such situations.

2.4.3. *Smurfing*

The *smurfing* or reflection attack, is classified as a Denial of Service (DOS). It consists of the attacker sending ICMP *echo request* queries to the *broadcast* address of a network, forging ("spoofing") its source IP address with that of the victim. All responses (ICMP *echo reply* packets) will be sent to the destination IP address of the target, causing saturation of the bandwidth or the IP resources of the victim.

However, the effectiveness of such an attack and its gain factor depend mainly on the network equipment that authorize (or not) the relay (forwarding) of broadcast packets. This being the case, a smurfing attack fulfills its potential for harm, because "spoofed" packets can reach networks other than the victim's original network. Otherwise, the impact of the attack is limited.

Pieces of network equipment that have recently been produced typically deactivate the *forwarding* of *broadcast* by default.

2.4.4. *Log Flood*

This technique aims to overwhelm the system logs of a computer, in order to make it unavailable. To do this the attacker will generate multiple events that will be recorded in the log files on the target system. This will then increase the size of the log files until they occupy all the free disk space. This may result in a malfunction of the affected system, because it will not have enough disk space for any other files.

This explains why the best practices for configuring a system involve's:

– the use of circular-type logs: a maximum number of records that the log file can accept, or the maximum size of the log file is defined in advance;

– and/or the offloading of system logs to a dedicated server with large storage capacity;

– and a hard disk partition partly dedicated to the logs. Saturation of log files will therefore not impact the entire system and avoid a malfunction.

2.4.5. *Worms*

Worms are part of the family of attacks on availability, on account of their intrinsic characteristic of diffusion. A worm is a program that replicates using the network capabilities at its disposal. The worm can then cause paralysis of a network, via saturation of links.

The first example was the November 2, 1988 Morris worm (named after its creator Robert T. Morris) which caused paralysis of almost the entirety of the Internet network for three days. Based on four software vulnerabilities existing in Unix systems, the worm diffused itself to around 10% of the machines connected to the Internet which, at that time, totaled 60,000 computers. Being able to copy itself several times on the same machine, the code paralyzed the network by saturating the bandwidth.

If the Morris worm was one of the first of its kind, it was certainly not the last. In May 2000, an email with the subject "I love you" and containing an attachment, began to spread over the Internet.

The scripts contained in the attachment, when executed, destroyed all image and sound files present on the victim's PC's hard disk, then consults the address book of the Outlook application, and diffuses itself by sending emails to all contacts. According to the study of Computer Economics, infection led to 6.7 million dollars in losses, and affected, according to ICSA.net[8], 65 % of American business with more than 200 employees,

Generally, a worm consists of two parts:

– the first is responsible for proliferation;

8 American company specializing in Internet security.

– the second is the active payload which aims to perform the tasks for which the worm was created (denial of service, spam etc.)

The principal components which have an effect at the level of system availability are situated in the proliferation part, except if the active charge is intended to contribute to denial of service attacks.

The world of "smartphones" is now no longer spared the problems of *malware* such as viruses and worms, because hackers have realized the "commercial" potential of these terminals.

Method of proliferation	Examples
Electronic message (email and MMS)	"I love you".
Instant messaging	Bropia.e (it suggested that you view a humorous image of a chicken with bikini suntan lines).
Network (Ethernet, Bluetooth, etc.)	Code Red or Nimda depend on vulnerabilities in IIS Web servers.
Surfing on an infected website	One method of propagation of the Nimda worm was based on a security breach named "Unicode Web Traversal exploit" that allowed it to install automatically and without the user's knowledge on a computer, when it surfed a Website that had previously been infected.
File sharing network (P2P)	The Kazaa worm filed a copy of its code in a sharing directory, in order to spread.

Table 2.6. *Example of principal modes of worm propagation*

Figure 2.13 demonstrates, that they actively follow the evolution of the market shares of different platforms (Android, iOS Apple, RIM BlackBerry, etc.) in order to develop their malware on the most commonly used systems.

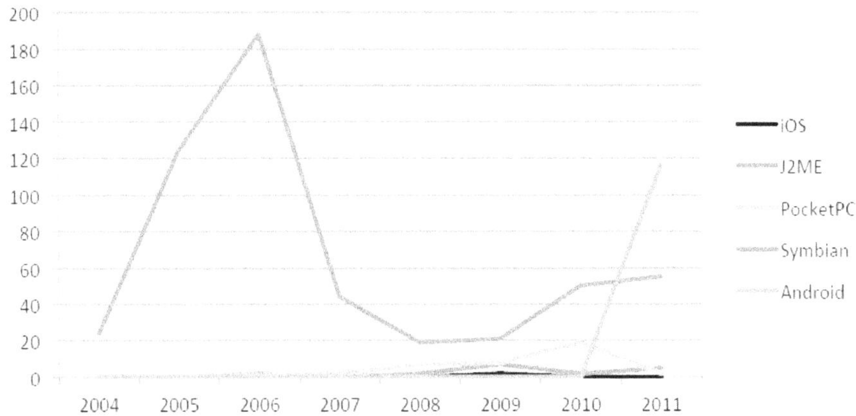

Figure 2.13. *Evolution of number of malwares by mobile platform (Source: F-secure quarterly report 2011)*

2.5. Attack on software integrity

A software virus acts in an almost identical manner to its biological counterpart; it therefore requires:

– a body, i.e. programs in which the virus will reproduce;

– a method of infection;

– a period of incubation before activating and carrying out the operation for which it was conceived.

These objectives depend on the imagination of the designer, and the virus can range from outright destruction of the system, via deleting files, reformatting the hard disk, to political messages such as the "Quarters" virus, which broadcast a message denouncing actions concerning the fight against illegal immigration by the British Prime Minister, and tried to saturate the Website of 10 Downing Street.

During the first quarter of 2012, the number of malicious pieces of software was over 80 million, while there were only 18 kinds of viruses in 1989 (source: McAfee).

And this exponential increase in the number of viruses is unlikely to slow, because the economic stakes have become enormous.

One classification of a virus was given by the SANS Institute.

On the basis of this classification (see Figure 2.14), viruses can be classified into four categories:

– *interaction in memory*: in this first case the virus is loaded into the RAM of the computer and infects any target program loaded into the memory. In the second case, the virus seeks to infect files on the hard disk;

– *action mode*: the viral charge may or may not be destructive, or may be a *dropper* (a virus which removes another virus);

– *dissimulation mode*: to escape antivirus and detection systems. It is here most notably where viruses are found which:

- attack antivirus systems;

- change form (polymorphs) to avoid detection by antivirus software. These are then forced to implement more complex techniques known as "heuristics" in an attempt to detect them;

– diffusion mode:

- "compiled": the virus is executed directly by the operating system of the computer. Here *boot sector* viruses[9] are most notably found;

- "interpreted": the virus is executed by an application. This is typically the case for macro-viruses written in the language of office software suits such as Word or Excel, or in messaging tools such as Outlook;

- multipart: viruses which use a combination of the compiled and interpreted modes.

9 The boot sector is a very specific area on the hard disk of all operating systems, which provide them the instructions to execute for system startup.

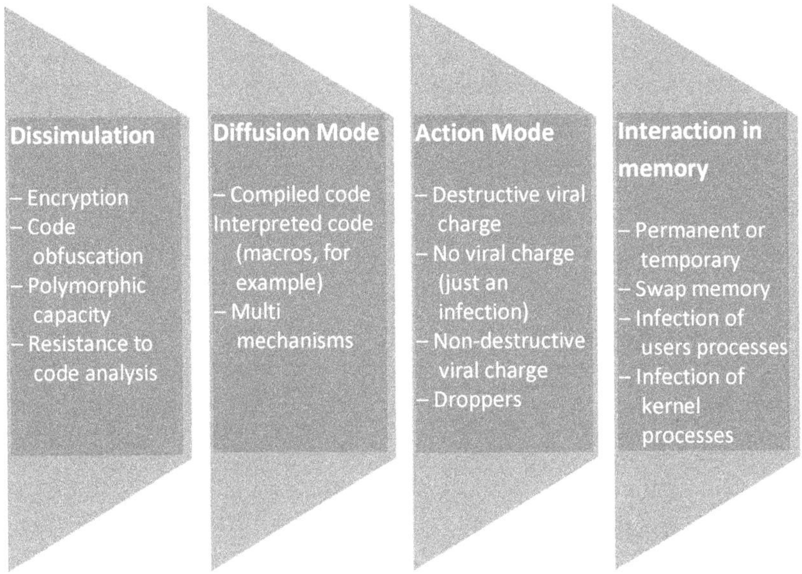

Figure 2.14. *Virus classification (based on the SANS Institute approach)*

2.6. BYOD: mixed-genre threats and attacks

Bring Your Own Device (BYOD) is defined by the fact that a company's employees use their own IT equipment for work. This includes the use of a telephone, smartphones, tablets, laptops or desktop computers. This movement, which began in the USA, has its origins in:

– the desire to reduce the cost of equipment made available to employees. It has grown significantly due to the enthusiasm of employees themselves in taking up this opportunity;

– the desire to improve staff productivity by enabling them to work anywhere.

These two causes were confirmed by a study conducted by BT during 2011/2012 on an international panel of 2,000 employees and managers across 11 countries.

Sample	Question	Answers	Total	UK	France	Germany	Spain	Italy	Benelux	USA	Brazil	China	India	Singapore
IT	To what extent would you estimate that the organizations that have adopted a BYOD policy benefit from a competitive advantage relative to those who have not adopted such a policy?	Percentage of those agreeing that there is an advantage	84%	74%	71%	79%	88%	78%	78%	82%	91%	98%	93%	91%
		To a certain extent	57%	51%	52%	65%	42%	62%	56%	54%	55%	74%	53%	69%
		It is a significant advantage	27%	24%	19%	14%	46%	15%	22%	28%	35%	24%	40%	22%
IT	When would you implement a policy allowing your employees to use their personal devices at work?	Percentage of those already implementing a BYOD policy, or anticipating to do it in the next 24 months	81%	76%	71%	69%	80%	78%	78%	83%	87%	96%	86%	91%
		We already have such a policy	40%	31%	39%	34%	39%	25%	36%	50%	51%	53%	38%	47%
		We will be implementing such a policy in the next 24 months	41%	46%	32%	35%	41%	53%	42%	33%	36%	44%	48%	44%
Employees	Does your employer allow you to connect to your organization's network with your personal device so that you may use it for work?	Yes	60%	37%	44%	50%	62%	63%	46%	52%	66%	92%	80%	62%
IT (respondents who have a BYOD policy, or plan to put one in place)	What benefits have you experienced (or do you anticipate) following the implementation of a BYOD policy?	Allowing employees to be more productive	64%	63%	61%	54%	75%	45%	62%	61%	66%	80%	60%	78%
		Offering employees more flexibility	48%	59%	45%	34%	46%	42%	65%	37%	46%	56%	56%	49%
		Allowing employees to better assist their clients	47%	47%	39%	42%	50%	43%	32%	47%	35%	61%	65%	46%
Employees (respondents who use peripherals bought personally for professional purposes)	What are the benefits of using devices bought personally for professional purposes (whether for yourself or for your employer?)	I am more effective and more productive	42%	34%	28%	44%	45%	34%	36%	47%	34%	52%	51%	53%
IT	In your opinion, are employees generally aware of the information security risk to the business in using a personal device in a work context?	Yes, all employees are aware of the risks	11%	4%	13%	7%	12%	4%	7%	9%	13%	20%	21%	13%
IT	Do you think that the majority of users understand access /permission policies linked to their mobile devices?	Yes – all users	19%	14%	13%	24%	21%	20%	18%	21%	19%	18%	28%	9%
Employees (respondents who use their personal device as their main or secondary work tool)	How would you evaluate the risk for the business in using a personal device in a professional context?	No risk	32%	33%	34%	32%	32%	40%	33%	36%	32%	26%	40%	10%
		A significant risk	25%	30%	12%	25%	24%	17%	17%	28%	29%	32%	19%	36%
IT	Have you encountered security vulnerabilities due to employees using unauthorized personal devices?	Yes	39%	36%	24%	35%	20%	38%	33%	31%	49%	40%	73%	58%
IT	Is allowing mobile workers 24/7 access to the business' systems now the biggest threat to the security of the IT system of the company?	All those who consider this to be true	83%	76%	79%	79%	71%	72%	89%	93%	84%	92%	89%	91%
		True. This change represents one of the greatest challenges for system security	28%	27%	20%	24%	18%	24%	18%	33%	36%	31%	39%	36%
		Partially true. This represents a new dimension in the manner in which we evaluate and put in place our security policies	55%	49%	59%	55%	53%	48%	71%	60%	47%	61%	51%	56%
IT (respondents who have a BYOD policy in place)	Which problems were you confronted with before authorizing your personnel to use their own devices in a professional context?	Security problems (malware, viruses, etc.)	74%	69%	67%	62%	73%	67%	75%	67%	72%	89%	78%	90%

Table 2.7. *Main questions and results of the BT enquiry into BYOD*

The flipside of BYOD is the potential opening of major breaches in the security of your information systems, if the risks are not properly taken into account and understood. The study by BT highlights several important points in terms of security:

– staff are aware of the benefits but not the risks: 42% of people who use their own equipment consider themselves more efficient and productive, while 1 in 3 perceives no risk;

– this lack of perception of risk is largely recognized by IT managers, since only 11% believe that users are aware of the risks of BYOD;

– the same study highlights that 68% of IT managers consider information leakage (theft or loss) as the greatest threat to their information systems.

The other main conclusions are summarized in Table 2.7 and Figures 2.16 and 2.17.

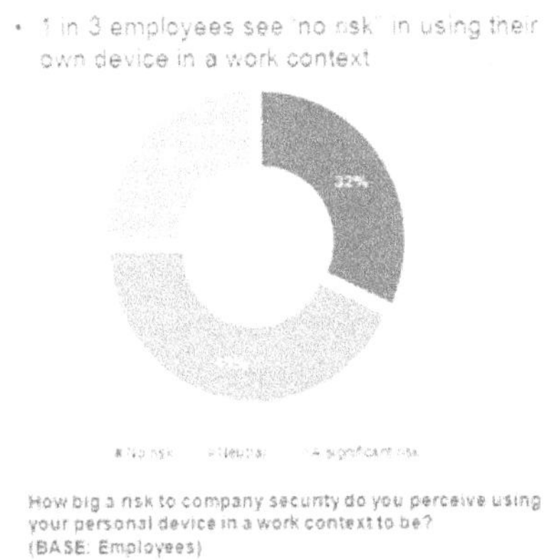

Figure 2.15. *Distribution of people perceiving no risk in using their own devices for work purposes*

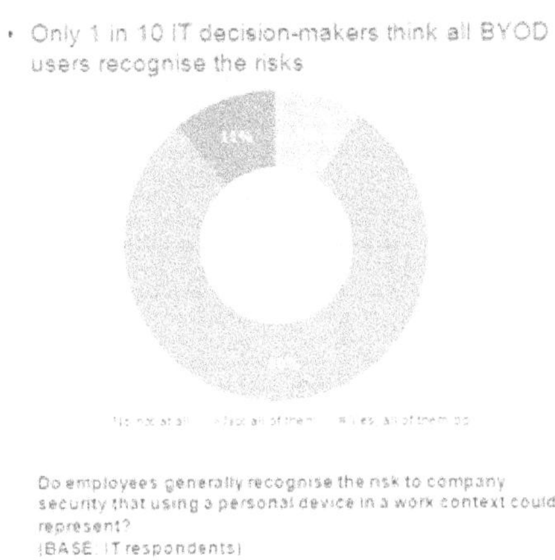

Figure 2.16. *Distribution of IT decision makers who are aware that staff underestimate the risks of BYOD*

Sample	Question	Answers shown on PPT	Total
IT (Base: Asked of respondents who have a BYO policy in place)	What effect has providing the security infrastructure to support a BYOD policy had? Combining the totals of those who rated '4' and '5 - strongly agree'	It's increased complexity	70%
		It's increased cost	60%
		It's changed how we prioritised projects	54%
		It's affected resource allocation	53%
		It's compromised resilience	45%
IT (Base: Asked of respondents who have a BYO policy in place)	What has been the effect on costs of your company's BYO program?	There's been a net cost increase	31%
		There's been a net saving	36%
IT	Do you perceive a BYO policy to impact or threaten your business's compliance or auditing obligations?	Yes	47%
IT	Have you experienced security breaches due to people bringing in unauthorised devices?	Yes	39%
IT	Are you able to detect if someone is using their own device on your network in a work context?	Yes and we are monitoring for it	43%
IT	Are you able to tell if an authorised user is using their device in an unauthorised way?	Yes immediately	33%
IT	Can you tell if someone is using an unauthorised device on the system?	Yes immediately	42%

Table 2.8. *Extract from the results of the BT inquiry about BYOD (Source: British Telecom "Rethink the Risk Research: international comparison May 2012")*

The BYOD movement is seen as a source of complication to the security of IT infrastructures, but also as a threat by the IT employees who responded to questions from the BT survey[10].

Although considered to bring cost reductions, BYOD also leads to many vulnerabilities that can weaken your system:

– due to theft or loss, potential information leakage of sensitive data is increased. In a study conducted in March 2012 by Symantec in the United States, 50 iPhones containing dummy data were deliberately lost. Symantec installed software on each of these smartphones to monitor what information was accessed by those who found the device.

Type of information	**% of people who accessed the dummy data**
Contacts	81%
Documents stored online in the *cloud*	47%
Social networks	64%
Passwords	57%
Salary information	45%
Online banking	43%

Table 2.9. *Priority information consulted on a lost smartphone*

10 British Telecom.

The results are explicit: between 40 and 60% of people accessed the data contained in the smartphone. This confirms the threat to information stored in these devices in the case of theft or loss:

– outside the control of your IT department (especially for smartphones). Sources of infection and infiltration are increased;

– by the significant mixing of personal and professional data. The risks to privacy and various litigations are thereby increased. Is software installed by an organization to find a lost or stolen smartphone fully compatible with the employee's right to privacy?

We will see in Chapter 4 how to put in place defenses against the inherent threats of BYOD.

2.7. Interception of GSM/GPRS/EDGE communications

Listening to calls, or the data exchanges of mobile systems, is often considered acceptable if one belongs to government or military agencies. However, other than images or ideas conveyed by various films or cultural tropes, interception of mobile systems has evolved to no longer be the exclusive preserve of governments or military agencies.

Modern mobile systems use GSM (Global System for Mobile communications) to transmit calls. With the evolution of consumer needs, particularly with respect to data, the standard has evolved to use GPRS (General Packet Radio Services) and EDGE (Enhanced Data rates for GSM Evolution), which is itself an evolution of GPRS.

Figure 2.17 presents the respective simplified architectures for GSM and GPRS. In both cases, the mobile system connects to *Base Stations* (BS), which are antennae distributed over a territory. These are connected to *Base Station Controllers* (BSC), which control the BS. In turn, the BSCs are connected and controlled by *Mobile services Switching Centers* (MSC). With the evolution into GPRS two new types of equipment were introduced: SGSN (*Serving GPRS Support Node*) and GGSN (*Gateway GPRS Support Node*).

In both cases, the GSM or GPRS/EDGE communications are subject to encryption. Three algorithms currently exist. The first, named A5/1 is in use in Europe and the United States; A5/2 is a variant used in other territories[11] and A5/3 is an improvement which will eventually replace A5/1.

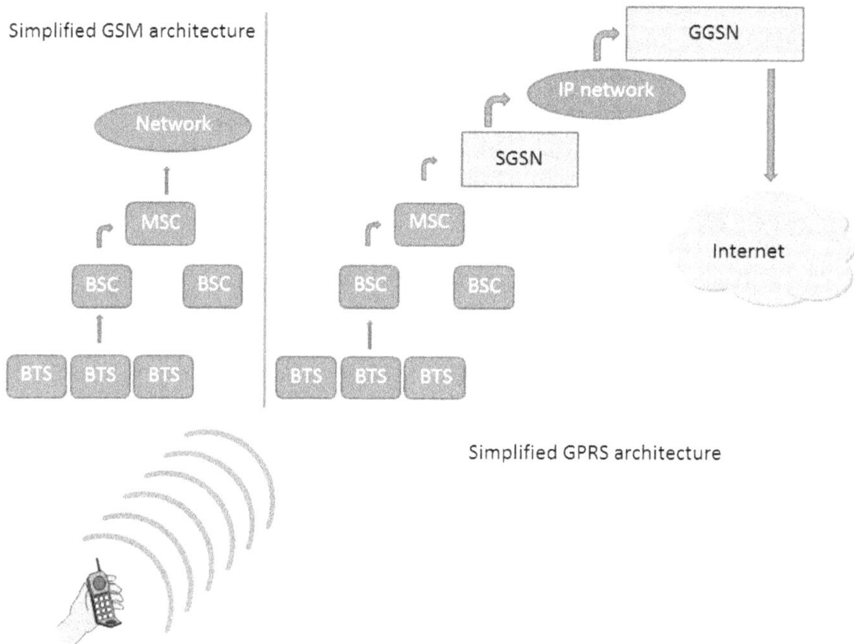

Figure 2.17. *Simplified diagram of GSM and GPRS architectures*

To date, numerous research projects have been conducted to assess the level of security of these algorithms, that were not initially publicly available[12, 13]. While many flaws have been detected and demonstrated in A5/1, the opportunity for practical attacks or interception was not in the public domain. However, during the last four years, the situation has changed due to:

---

11 Various work on this variant showed that it could be easily cracked: *Instant Ciphertext-Only Cryptanalysis of GSM Encrypted Communication*, E. Barkan, E. Biham, N. Keller, 2006.

12 M. Briceno, I. Goldberg, D. Wagner, *A pedagogical implementation of the GSM A5/1 and A5/2 "voiceprivacy" encryption algorithms*, //cryptome.org/gsm-a512.htm (www.scard.org), 1999.

13 A5/1 was developed in 1987; after certain leaks in 1994 its complete architecture was discovered by *Reverse engineering* in 1999 by Briceno.

– on the one hand, the development of certain *open source* tools that can create/simulate Bases Stations (BS) and controllers (BSC)[14];

– on the other hand, on the basis of research which led to the creation of *rainbow tables* to crack the A5/1 GSM encryption[15].

The interception of telephone communications at the GSM layer has therefore been made possible. In July 2010, at the DEF CON 18 conference, Chris Paget gave a public demonstration, and moreover, without even having to crack the A5/1 encryption:

– using a false *Base Station* to emit a stronger signal, Chris Paget was able to force certain telephones to use his own *Base Station*[16];

– then, taking advantage of the GSM protocol design itself, he requested that the captive telephones deactivate their ability to encrypt communications. Normally such a deactivation should generate an alert on the user's terminal, but for "convenience" this functionality has not been implemented by the various operators;

– therefore, the calls transmitted via captive mobiles could be recorded. In his *Proof of Concept* (POC), outgoing calls from captive telephones were then redirected to a voice over IP system so that they could be transmitted.

Using an approach similar to that of the *Blackhat 2011* conference, another attack was made public which captured DATA communications, by establishing a *Man in the middle* between the phone and the computer of a potential hacker. In the proximity of the target's phone, a false *Base Station* (BS) transmits a stronger signal than a legitimate BS. The target phone then communicates with an illegitimate BS, which asks it to disable its encryption layer. From that moment the telephone's traffic is diverted to the rogue devices; the pirate then records the data traffic of their victim.

So that the target is not using the GPRS protocol, which would make the attack currently difficult, interference is directed to the target smartphone on the frequencies used by the GPRS, forcing the target smartphone to only use the GPRS/EDGE layers.

---

14 Open BSC and Open BTS.

15 Project Kraken: srlabs.de/decrypting_gsm/ based on the work of KarsenNohl.

16 There is no mutual authentication between the *Base Station* and the mobile terminal; only the telephone authenticates itself by sending its IMSI (*International Mobile Subscriber Identity*).

As can be noted, interception of voice traffic and particularly data are no longer the preserve of the few government agencies. It is now a threat that businesses should consider.

Since antiquity, every invention of a new weapon or a new tactic has always seen the development of a countermeasure to limit its effects. Thus, the shield was created to parry the blows of the sword, and the radar was developed in order to identify aircraft in flight. It's the same with computers, as discussed in Chapter 3: for each method of attack that we have presented to you, there is at least one defense method.

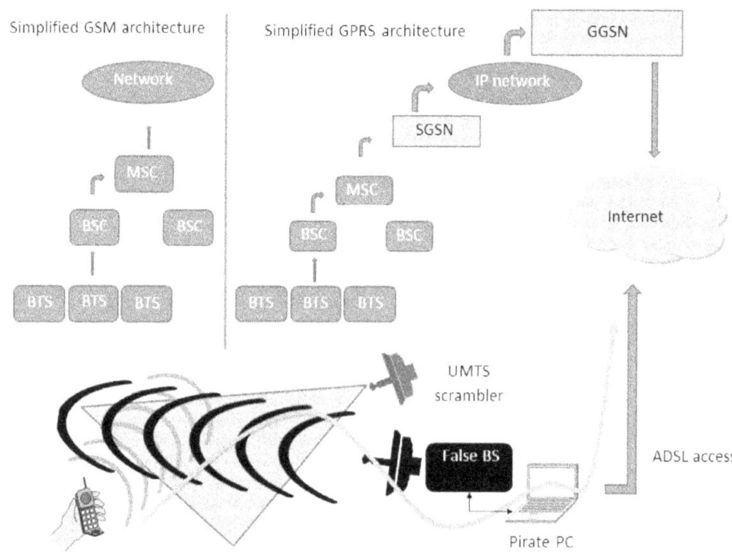

Figure 2.18. *Man in the middle attack for data traffic on the GPRS/EDGE layer*

Thus was created a market where vendors and consulting firms sell their services to prevent attacks from hackers and where, paradoxically, their experts are sometimes repentant former *hackers*. We are witnessing the creation of an "eco-system" operating in an autarky in which former criminals become vigilantes; because who has a better understanding of the techniques and practices?

Chapter 3

Technological Countermeasures

> "The unexpected happens. You'd better prepare for it"
>
> Margaret Thatcher
> (Prime Minister of the United Kingdom 1979-1990)

Since Antiquity, the defense models developed to offer protection from a potential attacker have changed little. Although techniques of attack and defense in the real or virtual world have continued to improve, the principles remain the same. For example, a city began risk prevention by identifying breaches and other weaknesses that could be exploited by an attacker and remedied them. Then the city installed sentries to detect an attempted attack and quickly raise the alarm. Thus, the city could implement at the earliest possible moment the appropriate means to counteract the attack and prevent it spreading to the entire community.

Today, the means of protecting modern information systems still rely on:

– prevention;

– detection;

– response.

After presenting the three major families of methods of protection, we will explore more specifically in Chapter 4 the case of security tools used in the context of mobility. Because of their intrinsic properties (the main characteristic being that they are not confined, unlike other components of an information system, to the physical perimeter of the company) mobile devices can only partially be serviced by traditional security tools. This model of protection (prevention, detection, response) must also be understood in the context of three critical ideas:

– no defense is impenetrable. This is something that should never be forgotten in the implementation and the consequent confidence that can be given to the protection of its information system;

– the loss, or breach, of a protection layer should not result in complete annihilation of the safety structure. This point is principally reflected in the model of deep defense as seen in the military and nuclear security domains;

– the protection model must delay as far as possible the attacker from reaching his ultimate goal, while allowing the defense to detect him as quickly as possible to be able to launch countermeasures. This is the principal idea of this model of protection, mainly based on the two points mentioned above.

3.1. Prevention

While computers connected to an organization's network are protected by the security systems and procedures of the company, this is not always the case for devices which access the information system remotely.

What methods should be implemented to ensure that company equipment (computer, PDA, etc.) is not infected or compromised during its time outside the company premises?

How is it possible to ensure that business data accessed by an employee at home or in a cybercafé, using equipment outside the company's control, will not be disclosed to a third party?

The prevention phase must therefore take into account the mobile device itself as well as the data it may need to access:

– protection of equipment will mainly consist of management of the vulnerabilities that can affect it;

– protection of data that the device could carry will implement cryptographic measures.

3.1.1. *Protection of mobile equipment*

3.1.1.1. *Management and administration of patches*

Bugs represent one of the first entry points for hackers who, on the lookout for any vulnerability, exploit them increasingly quickly:

– 2003: six months between the appearance of a vulnerability and the Slammer worm;

– 2004: 17 days to exploit a vulnerability by the Sasser worm;

– 2005: four days between the date of publication of the Microsoft MS05-39 vulnerability and the appearance of the Zotob worm which exploited it.

In addition, zero-days have become more and more numerous: this concerns attacks which exploit a security vulnerability discovered by hackers. Kept secret by the attackers, these zero-day vulnerabilities are among the most difficult to overcome, because the software vendor must deal with the problem urgently, and distribute patches as quickly as possible.

It is therefore critical to carefully manage the installation of patches on an information system, and in particular on mobile devices, because they are more exposed, not being protected by an organization's security systems (firewall, IPS, etc.).

To manage the deployment of patches, depending on the operating system's type there are various software solutions for fleet management, and for planning the distribution of *patches* classified according to their criticality in terms of safety. In all cases, the deployment methodology must follow six steps to avoid any possible damage to existing systems.

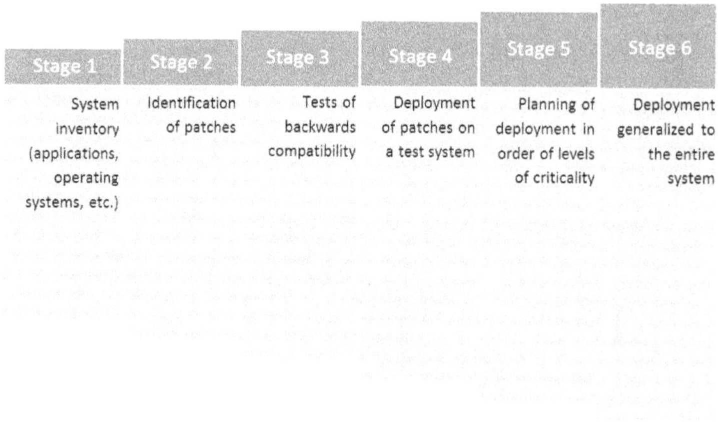

Figure 3.1. *Methodology for patch diffusion*

But in the case of mobile devices, an organization may decide to adapt this methodology to take into account their specificities. As these devices can remain unconnected to the company network for some time, and therefore unable to receive patches, if management of their distribution is centralized, it may be preferable to modify their configuration so they can download patches directly from software vendors' servers when they connect to the Internet.

In this case, different risks must be balanced: in this configuration, it is not possible to follow steps 2 through to 5, which may cause instability in the population of mobile devices (incompatibility between software installed on the device and the newly-patched operating system, etc.). In contrast, however, the latest developments in operating systems and software will be available, which can then be used to counter attacks attempting to exploit recently-discovered bugs.

3.1.1.2. *Locking the phone*

To prevent users from installing software directly downloaded from the Internet (freeware, pirate software, etc.) or change the configuration settings of the operating system (security level of the firewall, etc.), it is recommended that the locking features policy included in Windows are activated. In this way, the system administrator can prevent tampering with the standard configuration of these computers, which reduces failure rates, since users can no longer "hack" their PCs.

In the case of smartphones and tablets, MDM (Mobile Device Management) solutions allow the definition of security policies which, when sent to mobile terminals, enforce the security principles you wish to apply to this type of equipment. It is these strategies which define which applications may or may not be installed on the terminal.

3.1.1.3. *Mobile Device Management*

Mobile Device Management (MDM) is an application that allows us to centrally manage a fleet of mobile equipment. During the last five years, mobile operating systems have evolved rapidly and extensively, as summarized in Figure 3.2.

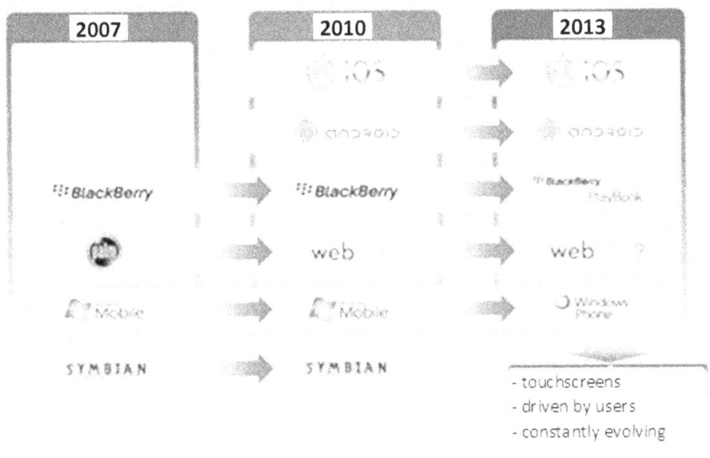

Figure 3.2. *Evolution of mobile operating systems*

At the same time, the definition of MDM has expanded to reflect the evolution of mobile OS from basic management (inventory tracking, basic configuration, locking of devices in the case of loss), to a more strategic form of management which now includes:

– taking into account several mobile operating systems;

– integration of BYOD;

– the significant exposure of peripherals to security risks;

– the explosion of the number of applications, and the volume of data related to this type of equipment.

From a security point of view, the main reasons to implement an MDM solution are presented in Figure 3.3.

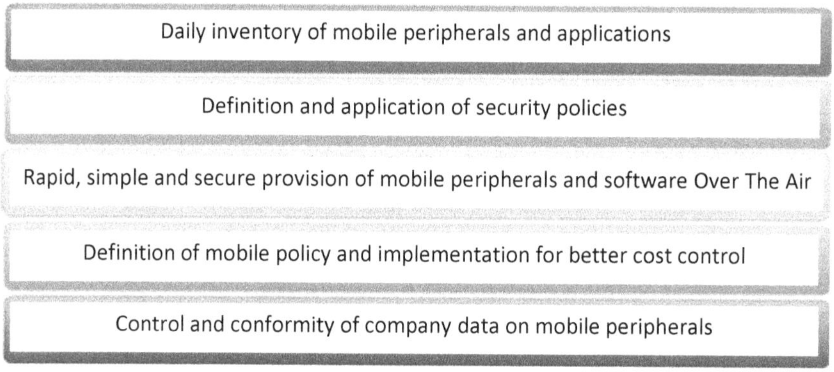

Figure 3.3. *Principal security justifications for an MDM solution*

3.1.1.4. *External storage peripherals*

USB ports offer very high transfer rates (480 Mbit/s for the USB 2.0 standard; 5000 Mbit/s for USB 3.0), and it is therefore much easier to copy large quantities of data in a few seconds. This is why a PC left unattended (in a cybercafé, an airport VIP lounge, etc.) can be "emptied" of all confidential information even when its owner is absent for only a few minutes. To prevent such a risk, it is therefore preferable to disable USB ports and use only CD/DVD readers on your organization's laptops. Although this method is the most effective, it is still the most difficult to implement given the resistance of users of the machines. Moreover, the current trend for laptops is to remove the CD/DVD to reduce weight, increase their battery autonomy and develop the USB facility as a storage unit!

In this context of strong reservations toward the application of such measures, a half-measure can be considered which offers a security guarantee which is certainly less reliable, but effective nonetheless: the disabling of the autorun facility when inserting any USB device; in other words, deactivating the reading and automatic execution of programs installed on a USB device.

3.1.1.5. *Screensaver*

There is a security tool which is often overlooked and yet is included by default in operating systems: the screensaver.

It does not just protect the computer screen with small animations (rain of stars, aquariums, etc.) to prevent it from being marked by characters displayed for too long; in particular, it also locks the computer after a certain period of inactivity, so that it cannot be used until authenticated again.

It is therefore strongly recommended to enable the screensaver with the automatic locking function to prevent over-curious passers-by from looking at your computer while you are absent, even for a moment, leaving it unattended.

3.1.2. *Data protection*

Any employee may need to use their mobile equipment to access confidential data (customer lists, manufacturing processes, marketing plan for the launch of a new product, etc.), which could harm the organization if disclosed or used by a competitor.

In an "ideal world" this should not be the case, as security requirements would take precedence over all other considerations, with an organization classifying its information assets and prohibiting all critical data at risk according to the risk of disclosure. But with the exception of a few companies and organizations with highly specific activities, such as those relating to armaments, most companies must take into account many other requirements (economic, organizational, etc.), and rarely apply such a policy.

This is why it is essential to implement all or part of the different security mechanisms that we present in this section.

A study by iBahn, dated October 2007, showed that the average laptop contained $525,000 worth of sensitive data. For example, let us recall some facts reported by the press:

– 2008: a laptop belonging to an HP employee in the Houston area is stolen, containing thousands of employee records (names, social security numbers, etc.);

– 2008: St George's Hospital in Tooting (UK) is robbed of six laptops containing data on more than 20,000 patients (name, date of birth, address, etc.);

– 2007: PointSec published a study showing that over a period of six months, over 100,000 lost portable devices (cellphones, PDAs, computers) were found in taxis in 11 cities (London, Sydney, Bombay, Stockholm, San Francisco, Washington, Helsinki, Frankfurt, Berlin, Munich , Oslo);

– 2006: a laptop of a Boeing employee was stolen, containing 382,000 employee records (name, social security number, etc.);

– 2006, the U.S. federal administration offers a reward of 50,000 dollars to whoever will return the "lost" laptop which contained more than 26 million records on its veterans.

NOTE.– It is equally important to carefully manage these mobile devices' lifecycles, because they may keep sensitive data on storage media or if one merely relies on a company that specializes in the recovery of old computers, this information can be found in the wrong hands

For example, in 2009, researchers from the BT security center acquired 300 hard drives on the online auction site eBay, which allowed them to determine that in 34% of cases, they contained personal or professional data. Among them, they were surprised to discover the security logs of the French Embassy in Germany, and information about how to counter Iraqi Scud missiles.

It is therefore essential to use specialized software to delete the data on mobile devices before throwing them away, selling them or giving them away. If you just remove the data via traditional means, information can still be recovered, by a technician, by directly accessing the storage media.

3.1.2.1. *Encryption of data stored on the hard disk or in flash memory*

In 2008, a study by the Gartner Special Report showed that only 25-35% of U.S. companies had implemented encryption tools on their laptops. Another study by the Government Accountability Office, also dating from 2008, showed that 70% of sensitive data stored on laptops of 24 major U.S. government agencies were also not protected by such tools.

The situation is the same for smartphones that have, for the most part, easily removable flash memory on which it is possible to store large amounts of information that should also be protected.

3.1.2.1.1. Encryption

At all times, the need to preserve the confidentiality of messages has been at the heart of the governments' concerns. Fear of interception of messages by enemies or opponents led to the development of methods, techniques and codes for preserving the confidentiality of exchanged information. Today, this is no longer limited to governmental, political or military spheres, as the need to protect information from the eyes of competitors, or simply safeguard private individuals' data, has become a more common concern.

The basic function of all encryption tools is to transform a message into a text which is incomprehensible to anyone not possessing the key (code) for reading it.

The oldest traces of these methods are found in Roman times. The famous Julius Caesar, in order to correspond with his generals, used as an algorithm the shifting of each letter by a number of predefined places relative to the usual order of the alphabet. With a shift of two letters, "A" becomes a "C" and so on.

Over time, encryption methods have been perfected alongside the development of those used for "codebreaking". Indeed, the stakes have increased: during World War II, breaking the code used by the German army[1] in the "Enigma" coding machine was at the heart of the Allied victory. At this point, encryption entered the era of widespread and now almost universal use of computers to help crack or encrypt messages.

Two main families of encryption methods exist today: methods known as symmetric and those called asymmetric. The first is based on the existence of a secret key shared by each of the recipients of the message, while the latter uses a public key known to everyone, but requires a second, secret key, known to the recipient of the message and to them alone. These two methods will be described in more detail in the remainder of this section, because they are currently the most commonly used.

1 Alan Mathison Turing, British mathematician, one of the fathers of computing, conducted research in the cryptoanalysis section, Hut 8, which succeeded in decoding the Enigma code.

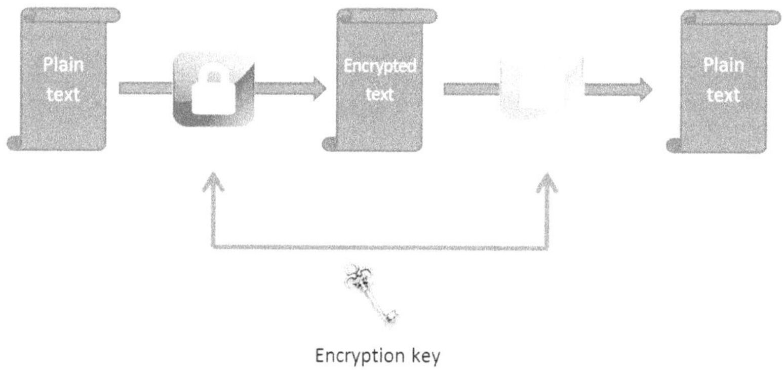

Figure 3.4. *Mechanism of symmetric encryption*

In the context of a symmetric encryption solution, a single key is used to encrypt and decrypt the message, as shown in Figure 3.4. This key, given its importance, must remain secret. Symmetric encryption is in fact called secret key encryption. The use of this method has the following features:

– the algorithms used are highly efficient and require less computing power than asymmetric encryption. This is the main reason for adopting this approach for embedded systems, for which the computing power and associated power consumption can be a limiting factor. Currently, taking into account the power of machines available on the market, this is less significant than ten years ago;

– symmetric encryption methods have two disadvantages:

a) on the one hand, the multiplication of keys as a function of the number of recipients; for N partners, $((N^2 - N)/2)$ keys will be required,

b) on the other hand, the transmission of the encryption key to the recipient must be carried out via a secure channel, a difficulty which increases exponentially as a function of the number of partners in a communication.

It immediately becomes clear that management of such a system, in the case of a large number of correspondents, can quickly become very complex. We return to this point later in the discussion of criteria for choosing an encryption method.

The best known algorithms based on the principle of symmetric encryption are so far:

– DES: Data Encryption Standard, created by IBM in 1973 and published as a standard in 1977. Because the key length does not exceed 56 bits, it is no longer recommended for use. An improved version, which is still in force, was subsequently developed: the triple DES;

– the RC2, RC4, RC5 family, created by RSA Security; the initials refer to the author of the algorithm, Ron Rivest;

– AES, or Advanced Encryption Standard became, after a competition in October 2000, the encryption standard for organizations of the Government of the United States.

For asymmetric encryption solutions, two keys exist (Figure 3.5):

– a first, so-called public[2] key is generated by the holder (correspondent A) and transmitted to the entity with whom they wish to communicate (correspondent B). No special precautions are applied with respect to this distribution;

– a second key, known as the private key, also exists, but in contrast to the public key, is never released, and is known only to its owner (correspondent A).

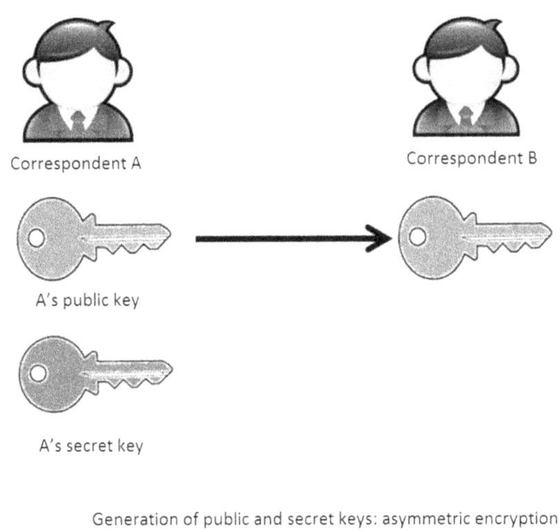

Figure 3.5. *The notion of public key and secret keys*

2 The key is called "public" because it is freely accessible and any entity may have access to it.

To send an encrypted message to A, party B will use A's public key to encrypt the information. From this point, A and A alone can decrypt the message using their private key.

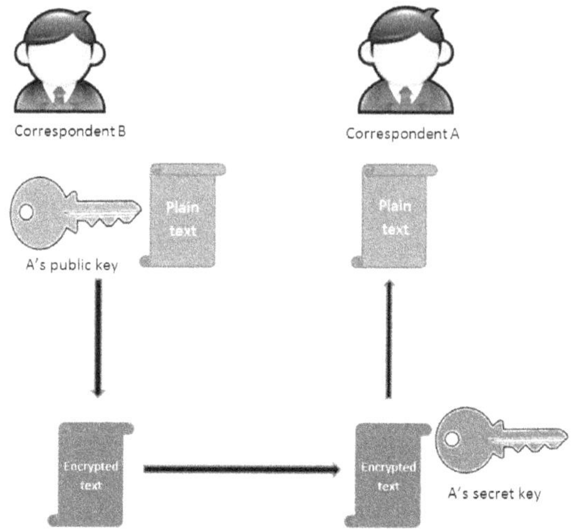

Figure 3.6. *Mechanism of asymmetric encryption*

One of the great advantages of the asymmetric encryption method is that it does not have to secure the transfer of the encryption key (public key) in order to encrypt the message to be sent.

Nevertheless, some constraints outweigh this advantage:

– firstly, this method involves a high consumption of computing power, which is not compatible with its use in some embedded systems. These asymmetric encryption algorithms are in fact based on complex mathematical functions that require many calculations and therefore monopolize the processor (CPU);

– secondly, the authenticity of the owner of the public key must be ensured. This raises the issue of the use of a trusted third party to certify that the public key of A is indeed that of correspondent A. In addition, the difficulty of revocation of public keys in the case of compromise must be considered.

The best known algorithms in this family are RSA[3] and DSA (*Digital Signature Algorithm*).

3.1.2.1.2. How to choose your encryption solution

The choice of an encryption solution is difficult if you are not well-versed in the field. Although there are many solutions on the market and in some cases encryption functionality is even included natively in an application or an operating system (Windows 7), it is imperative that certain criteria are considered before the final choice of a solution:

– never attempt to develop a "home"encryption algorithm; there are many in which the algorithm is well-proven, all of which have undergone many attempted attacks, and which can therefore legitimately be considered robust;

– follow good practice with respect to the use of an encryption product; you can use the best encryption solution on the market, and it will be of no use if the secret key encryption/decryption is stored in unsecured plain text.

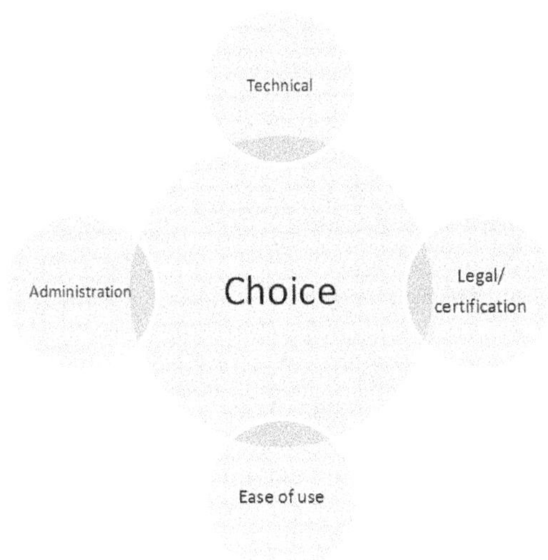

Figure 3.7. *Criteria for choice of an encryption solution*

3 Public key encryption algorithm whose name comes from the initials of the three researchers at MIT (Rivest, Shamir, Adleman) who developed it in 1977. It is based on the mathematical principle that it is easy to multiply two integers together (the result of which forms the public key) but there is no simple way to decompose a large number into a product of prime factors (forming the secret key).

In the second stage, the following specific criteria should be considered.

Technical

This consists of the choice of:

a) type of encryption (symmetric or asymmetric);

b) the encryption algorithm (AES, 3DES, etc.);

c) the length of the key (128 or 256 bits, etc.);

d) maximum processing time for encryption and decryption functions;

e) secure erase functionality (that is, irreversible).

Administration

a) This consists of considering recovery procedures, that is, the possibility for the product administrator to recover data in case of loss of the private key by the user. It is sometimes possible to use specialized key escrow companies, whose task it is to archive encryption keys, only revealing them to their owners or to a competent judicial authority, if so requested.

NOTE.– When implementing this type of solution, it is always important to provide a mechanism for the administrator to decrypt the data, because it may happen that the key is "lost" by the legitimate user, or an employee leaving the company "forgets" to return it before leaving.

b) The possibilities of access to a directory of public keys by users should be explored, thus avoiding difficulties in obtaining the keys of their correspondents.

Legal and certification

It is necessary to:

a) take into account the legal constraints that may be imposed in some countries, verifying that the permissions for use have been issued by the authorities of the state in which you wish to use the product;

b) check the level of certification (e.g. common criteria EAL4+[4]) obtained by the product, issued by a third party.

Ease of use

The following should be considered:

a) the possibility of using the application without requiring the installation of a fat client on the user's computer;

b) the possibility of using an auto-decryption function allowing an encrypted message to be read without having to install any software to do so;

c) possibilities for "on-the-fly" encryption and decryption in dedicated repositories, but in a manner transparent to the user.

NOTE.– Although many commercial software packages exist, it is important not to forget open source software, among which one of the best known is GnuPG. The attraction of such a solution, in addition to being free, is the availability of the source code which should, in theory, ensure that it is free from backdoors.

3.1.2.2. *Virtual Office and automatic computer cleaning*

There are two solutions for ensuring that the data used during a remote connection session cannot be read by a third party:

– the virtual office "emulates" a working environment for the user on their computer or on one of their organization's servers. Thus, all "footprints" that may have been left during data processing in this virtual environment will disappear automatically when it is deactivated;

– it is also possible to perform automatic cleaning of data that have been generated or downloaded to the workstation at the end of a remote login session.

4 The Common Criteria (or ISO 15408) allow certification of products and information systems according to seven levels of confidence (also known as levels of assurance) of increasing value (EAL1 to EAL7). They emerged from the convergence, in 1996, of the American (Orange Book) and European (ITSec) standards.

We will not develop these methods any further here, as we will return to them in Chapter 4.

NOTE.– There is another technique to avoid leaving traces of a connection using a publicly available computer, which also avoids contamination by the potential presence of malware. This consists of initializing (booting) it from external media such as a USB stick or a CD-ROM that contains an operating system and all the software needed to establish remote access. Thus, all data created during this connection are directly stored on this external media or in RAM, without ever accessing the hard disk of the computer. The user must simply retrieve their removable media at the end of the connection.

The main drawback of this method is that it requires that the publicly available computer is configured to boot from an external device, which is generally not authorized by the operators of these PCs.

3.1.2.3. *Remote destruction of data*

Some software vendors, MDM solution and some equipment (BlackBerry) offer the possibility of remotely initiating the order to destroy all data contained in a computer or a smartphone if it is lost or stolen.

If, in principle, this idea seems appealing, in practice its implementation is problematic, because it requires that the equipment is still connected to a network in order to receive this destruction order. It suffices therefore for an attacker to ensure that the machine is not connected to a network to commit his or her misdeeds.

3.1.2.4. *Screen filter*

To prevent an ill-intentioned individual reading the data displayed on the screen of your laptop when you work on it – which is easy to achieve, for example when you take the train – some companies, including 3M, have developed "secure" screen filters.

These reduce the viewing angle so that only someone seated directly in front of the screen can read its contents. Thus, any curious person seated to your right or left sees a blank screen.

3.2. Detection

The second layer of protection is to detect attack attempts as they occur, whether or not they manage to permeate the perimeter of the defenses (firewall, proxy, etc.) which have been put in place.

3.2.1. *Systems of intrusion detection*

3.2.1.1. *Networks sensor (NIDS/NIPS)*

In all military thinking, detecting the enemy as quickly as possible is one of the fundamental elements of a defense system. In the context of information systems, the principle is identical.

IDSs (Intrusion Detection Systems) are tools that analyze, at strategic points of your information system, the network traffic, generating alerts when a threat is detected.

The architecture of IDS has been subject to standardization efforts that led to the definition of the *Common Intrusion Detection Framework*. This revolves in a synthetic fashion around the five components depicted in Figure 3.8.

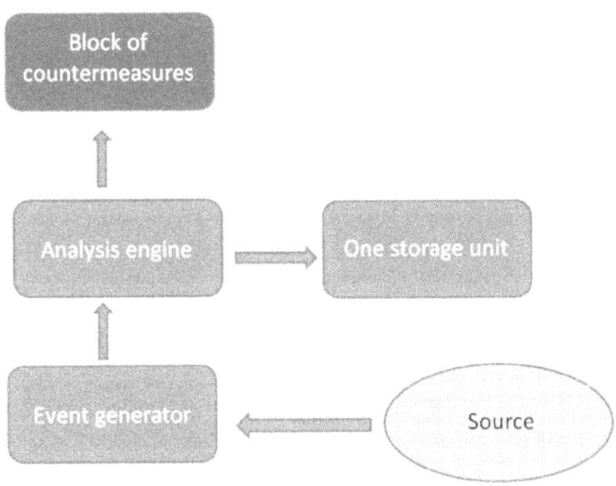

Figure 3.8. *Standard architecture of an IDS*

There are two broad families of IDS:

– NIDS (Network Intrusion Detection System), which focuses uniquely on the analysis of network traffic;

– NIPS (Network Intrusion Prevention System), which is an NIDS with an additional functionality allowing traffic judged to be anomalous/a threat to be blocked.

These tools can be:

– software to be deployed on dedicated servers, or on the machines to be monitored;

– an appliance, that is, a machine configured specifically by the vendor of the IDS solution for this purpose alone. Very often in this case, the configuration of the machine can modify the operating system to use only the bare minimum needed to run the IDS, in order to optimize performance;

– a dedicated card which is additionally inserted into the chassis of a router, a switch or a firewall.

Structured according to the architecture presented previously, IDS use a combination of detection modes to increase their effectiveness.

These detection methods are generally:

– pattern matching: this is found in groups of events with elements characteristic of a threat (e.g. Windows command line, Unix shell executions like /bin/sh, etc.). In all such cases, it is a static method that requires regular updating of the database of signatures based on evolution of attacks, when they are discovered;

– behavior analysis: here, based on statistics such as CPU utilization, the volume of emails exchanged, network traffic on certain times, etc., "normal" network or system behavior is defined. Deviation from this pattern of behavior will then trigger an alert;

– protocol analysis (principally for NIDS): packets that travel across the network are analyzed to verify their compliance with standards (RFC, etc.) in order to detect traces of attacks (e.g. any ping packet with a size of 65,535 bytes indicates an attempted ping of death attack).

In the course of acquiring this type of tool, it is necessary to analyze what the market offers based on four criteria depicted in Figure 3.9.

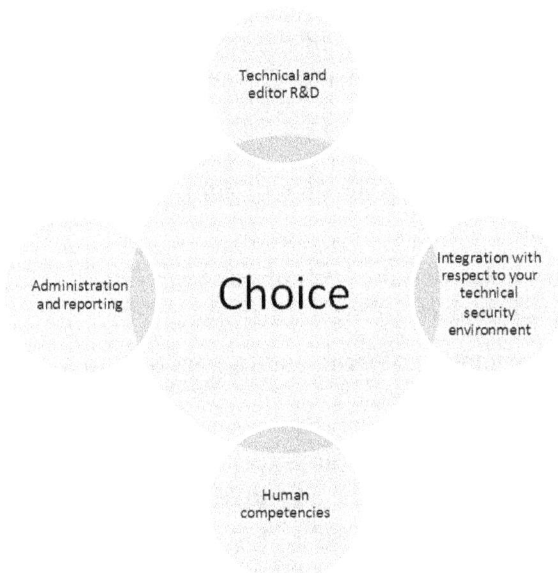

Figure 3.9. *Criteria for choice of an IDS solution*

Technical and editor R&D

The product must:

a) provide continual and rapid adaptation in response to changes in attacks developed by hackers. Significant R&D support must be one of the criteria taken into account when carrying out a technical analysis of a solution;

b) be able to analyze network protocols in-depth and decode the contents of the application layers;

c) be capable of handling very large volumes of data in both the collection and analysis phases as well as recording in its databases. A probe installed in front of the devices providing Internet access must be able to receive traffic of several Gigabits per second. The maximum number of packets processed per second without loss and the maximum number of concurrent sessions that can be managed, are taken into account;

d) have the ability to record suspicious traffic to allow *post hoc* analyses of hacking attempts to be performed and serve as evidence for possible legal action;

e) have access to some or all of the signatures used in order to properly analyze alerts and determine whether they are false alarms.

Administration and reporting

The product must allow generation of information-rich reports that can be adapted to suit your needs:

– definition of alerts by the level of risk (e.g. certain IP addresses can be more critical, such as those of the R&D department);

– automatic generation of trend analyses tailored to a particular audience (management, security expert, etc.).

A very good IDS without an excellent reporting module is useless because the tool must not only be able to detect attempted attacks but also to raise awareness of risk depending on the type of people for whom reports are destined.

Administration features should be considered in terms of:

– ease and costs in terms of training and use. A tool which is powerful but very complicated to use requires significant training costs, which may have a non-trivial impact if teams responsible for using the system experience periods of turnover. Otherwise, you may have a potential disruption in the daily operation of your IDS/IPS;

– the wealth of means of alert notifications available (email, fax, SMS, SNMP trap, auditory signal).

Integration with respect to your technical security environment

a) You must ensure that the selected tool can easily be interfaced with your organization's safety equipment. A NIDS/NIPS should be able to communicate with or understand the information it receives from other devices (firewalls, routers, etc.) to block attack attempts and initiate response measures. For this there are several standards, defined by the major market players including Active Security, ANSA, and OPSEC.

b) You may also explore the possibility of using agents installed on the systems to be monitored in a complementary manner (see section 3.2.1.2 – *Systems probe (HIDS / HIPS)*).

c) In addition, you must verify the ability of the tool to correlate the events generated by a single attack, but reported by different equipments.

Human competencies

The selected tool cannot be used effectively if the teams chosen to use it do not have sufficient technical knowledge to do so. IDS requires very good technical knowledge of protocols and potentials attacks that can be implemented, in order to spot false alarms in the list of events that is generated. The implementation of an IDS will necessarily involve a long phase of learning and improving the configuration of the detection system.

These improvements can be made only with the strong technical expertise of your staff, lest the system be underused or poorly configured, and hence failing in its main objective: to alert you when you are under attack!

NOTE.– Of course, apart from commercial products, there are many *open source* tools: Snort is the best known and also served as the basis for the development of several paid solutions.

NOTE.– If you wish to use solutions for auditing information system security (see section 4.4 – *Control of conformity of the VPN infrastructure*), it should be borne in mind that they will simulate attacks and network discoveries. It is therefore important to configure your IPS/IDS so that they do not generate alerts when these test attacks are launched.

Once you have selected the NIDS/NIPS to suit your needs, it is necessary to ensure its correct implementation. At this stage it is necessary to be aware of the limitations of these tools, to avoid a false sense of security, because there are several limitations to these tools:

– a NIDS/NIPS, without inline SSL decrypting functionality, does not scan encrypted streams that pass through your network. This is the downside to implementing encryption solutions. Thus, attacks that can be carried out via protocols using encryptions will not be analyzed if they are part of your legitimate traffic information system;

– an attacker can make their attempt not match the signature that is referenced in the IDS/IPS base. Here we can see the importance, as we described in the section on selection criteria, of being confident that the software vendor has an R&D team which can quickly adapt its products according to hackers' "innovations", and even anticipate them;

– because an IDS can generate many false alarms, it may happen that your operations teams become completely overwhelmed to such a point that they could be discouraged from continuing to use this tool. It is therefore vital to provide a *tuning* phase long enough to adapt the tool to the context of your business;

– do not make the mistake of considering the NIDS/NIPS tool in isolation, and not integrating it into a broader vision of a SOC (*Security Operation Center*) or SIEM (*Security Information Even tManagement*). When investing in a tool like NIDS/NIPS, it is important not to consider it as a simple alert tool, but as part of a larger SOC/SIEM architecture. We will return to this subject in more detail in the following sections.

3.2.1.2. *Systems probe (HIDS/HIPS)*

In order to protect servers and clients, you can use specific software called HIDS (Host-based Intrusion Detection System) and HIPS (Host-based Intrusion Prevention System) to detect and eliminate malicious acts. To do this, these tools typically use:

– a set of traditional methods based on research into the signatures of known attacks and on verification of file integrity:

a) concerning pattern matching (static research of specific structures denoting an attack); since this topic has already been discussed above we will not return to it further;

b) concerning verification of file integrity, this consists of taking an imprint of the files that are considered to be sensitive, which allows detection of any modification or deletion which would be evidence of an ongoing attack. This type of control, carried by the tripwire tool, for example, can go far beyond the simple verification of the presence of the file and monitoring changes in fields in the registry database of a Windows system;

– a set of two complementary strategies:

a) the first consists of an application control. It requires a prior declaration of applications authorized to run on the computer, as well as the resources (files, registry, network access, etc.) which they can use. This is therefore a similar approach to that of a firewall, but whose scope only affects the computer's applications and resources;

b) the second is based on behavioral analysis, where the aim is to detect any unusual action, some behavior different to that expected under what is considered the normal use of the computer. It is in particular during the learning phase that this set of rules can be defined. For example, the fact that an application uses new system calls might be an indication that it has been infected by a virus.

If HIDS/ HIPS offer advantages compared to their NIDS/NIPS counterparts, including their ability to analyze encrypted streams (SSL, etc.) since they exist in plain text at the level of the computer itself, they entail the same complexity in terms of their proper configuration in order for their alerts to be relevant. Thus, for example, monitoring of call systems of many applications existing on a computer can rapidly become a nightmare in terms of monitoring and administration for your security implementation teams.

It is therefore important to emphasize the use of such tools for monitoring particular events which denote a real threat. For this, it is necessary to:

– have a genuine expertise in the functioning of the given operating system;

– know which applications to monitor;

– have defined a computer compliance policy, in order to have a repository in which to refer in order to define the scope of control, and ultimately the alert level.

There are other tools which can detect intrusion attempts as early as possible, such as software for collecting and analyzing logs. We present these in section 3.2.3.1 – *Collection and analysis of events*, because their domain of application is much wider.

3.2.2. *Honeypot*

Although not a new concept, this was introduced for the first time by Clifford Stoll in the late 1980s. The term *honeypot* was invented in 2001 by Lance Spitzner, who gave the following definition: "a security resource whose value lies in being probed, attacked, or compromised". A *honeypot* is therefore, as the name suggests, a "pot of honey", whose function is not to attract bears but hackers. To achieve this, it will simulate one or more virtual machines, or even entire networks, and record all attempted attacks destined for these simulated computers.

Clearly, therefore, the *honeypot* approach is very different from other types of security tool generally implemented. The *honeypot* framework approaches an active technique insofar as one acts/interacts witht he attacker. It is therefore not at all a passive strategy like other security systems such as *firewalls* or antivirus software that "simply" waits for the attack to approach.

3.2.2.1. *Why implement a honeypot?*

The implementation of a *honeypot* fundamentally addresses the need of collecting information about the attacker, their methods, their goals and so on, by providing him with a clear target in a familiar terrain (the *honeypot*). Depending on the nature of the organization which decides to implement the *honeypot*, the goal will be different:

– in the context of a commercial structure, the *honeypot* addresses the need to protect;

– in the context of a research or governmental organization, the objective is to collect the maximum amount of information about the hackers, in order to develop plans of action for countering them, or to inform the community of the risks or methods they may encounter.

As you will have noticed, in both these cases, the *honeypot* addresses the needs of an increase in security levels, according to the methodology outlined in this book, namely:

– better prevention: understanding the methods of the attacker in advance helps to define the best strategy to counter it. The analysis of the modus operandi used by the hacker will eventually lead to the detection of new attack techniques and new vulnerabilities in operating systems or applications;

– advance detection: unlike IDS, a *honeypot* never generates false alarms, because, in principle, only a pirate attacking the *honeypot* generates an alert. By revealing the attempted attack in real-time, the *honeypot* improves the organization's defense capabilities;

– an accelerated reaction: as a consequence of the previous point, knowing about the attack, the defender can react faster and organize themselves to counter or reduce the impact. Moreover, by focusing on attacking the *honeypot*, the attacker loses valuable time, the "pot of honey" being, after all (whatever its technical complexity) nothing but a decoy, offering no information of value to the company.

To achieve these objectives, companies that implement this type of technology must respect the principles that we will outline in the following sections, lest they pose a greater risk to their information systems if they had not used a *honeypot* in the first place.

3.2.2.2. *How to choose your honeypot*

The main selection criteria for choosing a *honeypot* are shown in Figure 3.10:

– level of interactivity: this is without doubt the most important criterion, because it is this which will have a direct impact on the complexity of the *honeypot*. If you opt for a solution based on a simulator/emulator of services and not a true operating system, interactivity is low and the attacker's scope will be limited. Conversely, it will also reduce risk. It is thus necessary to define:

a) the maximum number of machines that can be completely or partially simulated (limited interactivity);

b) the type of operating systems and network equipment to be simulated;

c) the types of service offered;

d) complexity for the attacker, via the choice of vulnerabilities that we selected;

– alert method (email, simple console, SMS, etc.);

– storage capacity: this will be directly linked to the complexity of the *honeypot* selected; a fully interactive system can generate as much and even more logs than a production system. We will have to define if we want to, if this is required:

a) record all the names of all files to which the hacker has had access;

b) capture all network traffic related to the *honeypot*. Record all actions that have been performed by the attacker (system commands, etc.);

c) perform an upload of the logs to another machine to make sure they cannot be corrupted and are available for use for any subsequent police investigation;

– the administration console: this can range from a graphic interface accessible via a Web browser for solutions with limited interactivity, to tools allowing full management of several *honeypots*;

– traceability of actions: finally, it is necessary to examine the level of granularity of information that you wish to collect on the actions that the hacker will perform on the *honeypot*, as well as the data which relate directly to him (IP address, who is, DNS, trace route, etc.).

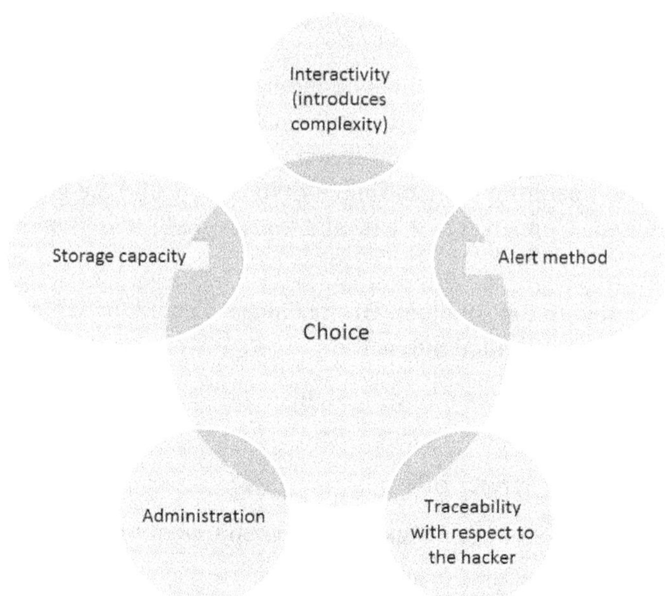

Figure 3.10. *Criteria for choice of a honeypot solution*

3.2.2.3. *Where should the honeypot be installed?*

Positioning is an important point in the implementation of a *honeypot*, especially if you plan to use a high interactivity system, because this means that the attacker will have greater room for maneuver in his actions and it is important to never forget that the *honeypot* is part of your information system. This means that bad positioning can introduce a major breach into your system.

The positioning of the *honeypot* will depend principally on the intended goal. The *honeypot* can be installed:

– at the Internet front end, if you wish to monitor attacks on the organization's important services (SMTP, DNS, etc.);

– amongst the production servers, if you wish to highlight a compromise of the internal network, or monitor unauthorized internal access or attempts to leak information by staff themselves.

Today, research into *honeypots* has made major advances and allows the simulation of complete networks, thanks to virtualization techniques of the network and system environments. But unfortunately, hackers have also learned to use these tools to trap the police investigating their illicit activities. The concepts of prevention/detection/reaction are also valid for the two camps.

3.2.3. *Management and supervision tools*

"The ability to correlate and evaluate thousands of security events generated by firewalls, antivirus and intrusion detection sensors is a major challenge for security administration faced by companies today. Organizations must evaluate real-time threats and possible consequences to distinguish events to be processed as a priority from those who have a low impact." Mark Nicolett, Vice President & Research Director, Security, Gartner Group.

3.2.3.1. *Collection and analysis of events*

In the context of security management, the collection of all the information generated by all the equipment participating in a safety architecture, and then their processing in order to use them in a relevant manner, have already become a major issue.

This is because it is in the logs (if equipment is properly configured) that it is possible to detect the early indications of an attack, or where initial research is carried out in the case of an incident. The collection process must:

– contain the maximum information. Whether analysis is in real-time or in batch mode it is imperative that the logs contain as a minimum the following information:

- session start and end date,

- status of the authentication phase (failure,validated, etc.) and identification of reasons leading to a refusal (invalid password, etc.),

- method of authentication used (OTP, etc.),

- identification of the initiating user,

- identification of remote equipment used (if this is managed by the enterprise),

- IP used/allocated,

- amount of data transferred during the session,

- type of VPN protocol (SSL, IPSEC, etc.) and version (3.0, 3.1, etc.) used and/or type of access (Web portal, tunnel, etc.),

- information gathered during the conformity analysis of the machine (type of operating system, version of the antivirus signatures database, etc.),

- applications/resources accessed during the session (IP address, DNS name, port number, URL, etc.) indicating the date of the request;

– be the subject of a regular backup onto offline media, to protect them from corruption on the part of the attacker;

– implement a mechanism to ensure evidence of the integrity of stored data, especially in the case of possible use for legal action. This device should also incorporate a mechanism for a trusted timestamp, a point which can prove extremely constraining;

– centralize these logs on a unique platform in order to later exploit them using a tool dedicated to alert correlation.

Centralization of logs of different systems into a single point, and their consolidation, greatly facilitates examining them, allowing cross-analysis. But in all cases, acquisition of a dedicated tool for analysis/alert correlation is necessary, because nowadays the daily volume of logs generated is too large to be usefully exploited without the aid of this type of software. This is the benefit of installing an SIEM, the subject of the next section.

3.2.3.2. *SIEM (Security Information Event Management)*

An SIEM is a device that aims to collect logs from various pieces of security equipment and in various formats (syslogs, logs owners, etc.), to offer an aggregated and real-time view of security.

To qualify as a SIEM, a system must operate according to the phases summarized in Figure 3.11.

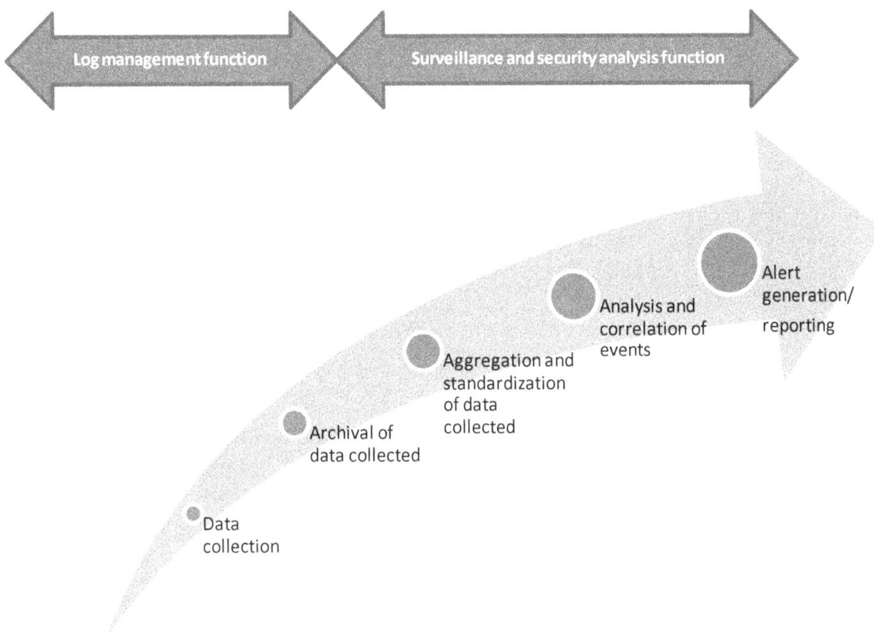

Figure 3.11. *Operational principles of an SIEM*

In existing organizations, the number of security devices, usually produced by different developers/vendors, has increased over time. In addition, such devices are not necessarily managed by the same team (the routers by the network department, *firewalls* by the security team, etc.). The need to bring together all alerts/events generated by the many and varied facilities resulted in the development of tools that can:

– provide the capacity for global processing of all data from a single place;

– correlate multiple events to detect intrusion attempts in the accumulation of logs, for example, an erroneous authentication error on a server in a protected area, outside opening hours and on a disabled account is likely to be a hacking attempt.

To provide this homogenous overview, the SIEM must complete five processing stages, as depicted in Figure 3.11.

Table 3.1 details the recommended processes for each stage.

Stage	Recommended processing
Data collection	Receive and interpret data from as equipment formats as possible (SNMP traps, proprietary *firewall* logs, syslog, etc.)
Archival of data collected	Archival is essential in the case of enquiries, and also if the collected data are to be presented to the judicial or regulatory authorities.
Aggregation and standardization of data collected	The format of logs generated by different tools is often proprietary, because this has not been the subject of a consultation between vendors. One of the tasks of the SIEM tool is therefore to aggregate them, which notably allows potential duplicates to be deleted.
Analysis and correlation of events	The collection of security events having by this stage been standardized, now information, indicators and traces constituting evidence of an attack must be detected.
Alert generation/reporting	Alert generation (sending an email, graphic visualization of an attack on a dedicated screen, etc.) based on a critical threshold and/or automatically implementing countermeasures to protect the information system.

Table 3.1. *Details of the five stages of operation of a SIEM*

NOTE.– The latest versions of SIEM support the IETF data exchange format for incidents: IODEF (*Incident Object Description and Exchange Format*) to foster collaboration between various community members involved in the fight against computer crime (TF-CSIRT).

NOTE.– So that alert messages cannot be intercepted and destroyed by possible hackers, certain organizations prefer to create a dedicated network, independent of that on which users' data is transmitted, to send management data as well as alerts.

3.3. Reaction

Once the detected threat is identified, the last level of protection (reaction) implements the appropriate methods for confining and eradicating it.

As we shall see later in this section, in order to improve efficiency some of these tools directly integrate the detection and response levels, such as *firewall* solutions, for example.

3.3.1. *Firewall*

A firewall is a safety device whose function is to filter the connections between networks. This filtering is traditionally carried out based on IP address source/destination and takes into account the services (ports) required. Based on the configuration provided, the firewall plays the role of a barrier, authorizing or denying access between networks that cross it.

In addition to the classic parameters of IP source/destination and service, depending on the technicality of the equipment, firewall rules can take into account other elements such as the timescale for opening a stream, prior user authentication via a directory, protocol type, etc.

Historically, the first firewalls had the single function of filtering network packets. Over the last 20 years, in line with the increased computing power as well as increased threat complexity, the functionality of the firewall has evolved (see section 3.3.1.1 – *How to choose a firewall*) as well as its nature itself within a safety architecture, sometimes ensuring (load balancing) functions of applications.

In addition, changes in the firewall were not only limited to its functionality, but also affected the equipment (hardware) on which it was deployed. In the 1990s, an installation was understood by the "hardening" of a server (Unix or Windows) on which the firewall software was then deployed. In terms of administration, in addition to the firewall itself, the operations team must also maintain updates to the OS server.

Period	Functionality
End of the 1980s	Network packet filtering with respect to source and destination IP address as well as port numbers.
Beginning of the 1990s	Filtering at the application layer, thanks to the integration of *proxies* allowing layer 7 traffic from certain applications (notably email and FTP) to be analyzed. This consequently allowed much finer-grained filtering of these applications (forbidding of the GET command in an FTP, rewriting of email headers, etc.).
Middle of the 1990s	Introduction of state full inspection filtering; the firewall became capable of remembering open connections, via the creation of state tables, which allowed it to follow the establishment of TCP/IP connections.
Beginning of the 2000s	Integration of IPS functionalities (see section 3.2.1.1), allowing detection of attack signatures and triggering adapted countermeasures.
Present day	Further application filtering (Web stream, SSL inspection, database flow, detection and mapping of applications transmitted over the network).

Table 3.2. *Evolution of firewalls over the last 20 years*

Over the past decade, firewall publishers have gradually evolved toward distribution of appliances (servers with OS pre-installed and optimized for the deployed firewall product) and removed "pure" software versions of their products from their catalogs. Although this solution has not always received a very positive reception from firewall users, it does increase simplicity, particularly because the vendor manages OS updates. In addition, the standardization of the operating system configurations simplifies support for vendors.

3.3.1.1. *How to choose a firewall*

To choose the best firewall for your needs, you should consider the five categories of criteria presented in Figure 3.12 and described in Table 3.3.

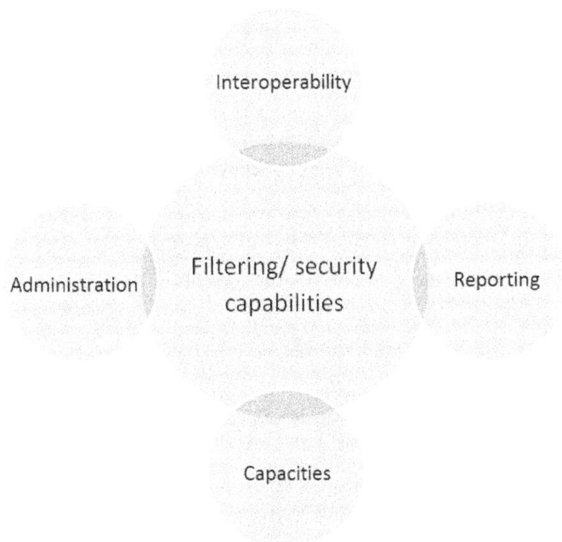

Figure 3.12. *Criteria for choice of a firewall*

Categories	Points to take into account
Filtering/security capabilities	"Stateful inspection" support Protocols supported (IPv4 and 6 etc.). Application filtering capability (SMTP, FTP, etc.). Minimal protection against DDoS (limitation of number of connections, detection and automatic rejection of incomplete TCP connections, etc.). Encryption, via management of VPN SSL or IPSEC tunnels. Integration of static or dynamic NAT. Proxy functions: HTTP, SMTP, etc. URL filtering, via public or commercial URL bases (for example to restrict access to pornographic content, games, etc.). Anti-poofing capacity (see section 2.2.2 – *IP poofing*). Date- or timetable-based filtering. Taking into account dynamic ports (SQLNet application, for example). Filtering of Voice-over-IP.

Administration	Has an advanced graphical user interface (GUI) that allows effective rule management and optimizes the cost of processing changes to equipment. Possibility for access for multiple users.
Interoperability	Interconnection with the corporate directory of users to manage rules which require authentication. Generation of alerts which are compatible with other correlation tools (IDS/IPS, etc.). Log format compatible with industry standards. Possible coupling with antivirus products. Ability to easily integrate with monitoring tools, via standard protocols (SNMP trap).
Capacities	Maximum number of communications that can be established. Maximum rate of information flow which can be processed. Maximum number of IP tunnels that can be created. Maximum throughput that can be achieved with an IP tunnel. Maximum number of users that can be managed. High availability architecture with the possibility of load balancing across multiple firewalls, which will withstand the scalability and provide greater fault tolerance. Dynamic allocation of IP addresses (DHCP function) to user workstations. Bandwidth module management (QOS) to prioritize the flow of certain applications deemed critical and thus avoid saturation problems.
Reporting	Dedicated report generation module. Automatic sending of reports. Personalization of reports according to requirements. Sending of alerts by visual message, SMS, email, etc. Log server independent from the firewalls themselves, to ensure processing of large amounts of data. Richness of logs generated: logs must provide sufficient information to allow subsequent coupling with SIEMS.

Table 3.3. *Principal points to take into account when choosing a firewall*

One of the points we have mentioned in the selection criteria "GUI with advanced administration functions" is often debated. For some users, it is more convenient and efficient to go through the command line to carry out daily tasks. For others, however, the administration of a fleet of hundreds of firewalls cannot be achieved without powerful graphical management interfaces (sharing of objects, networks, etc.). Without wishing to give a definitive answer, it is important that you take into account the "sensitivity" of your administrators, but also the level of maturity of the administration interface (graphics consoles with many bugs, etc.). It is essential to take this point seriously because, as stated by Jean-Marc Puigserver (product manager at Microsoft Security and Administration) in an article of 01 networks: "80% of firewall security problems stem from bad configuration". A powerful administration tool can significantly help to reduce these problems.

NOTE.– While you make your choice, you may also wish to take into account certifications (Common Criteria, ICSA, etc.) obtained by the various products, as these are a guarantee of quality.

NOTE.– It is preferable to install the firewall on a separate, dedicated machine, because each new piece of software installed can bring new security vulnerabilities and the firewall may need to use all the machine's resources (memory, hard drive, etc.) in the case of an attack.

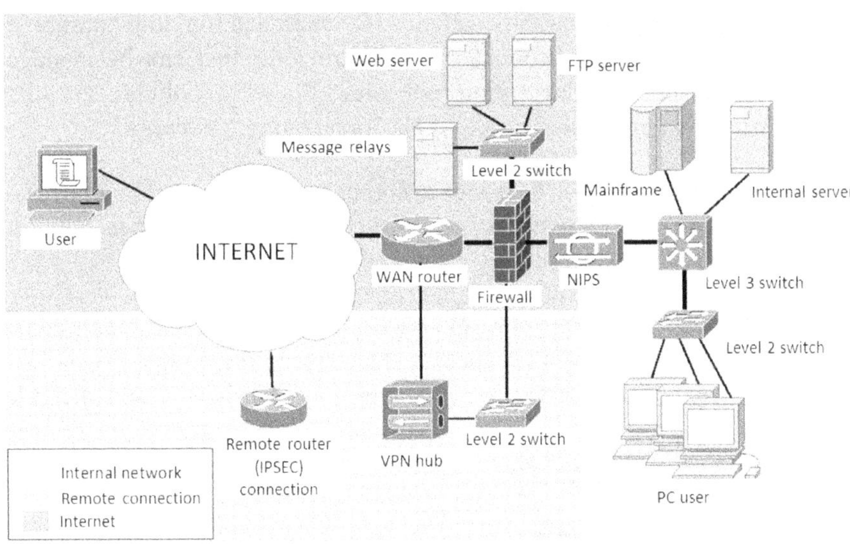

Figure 3.13. *Example of firewall implementation*

3.3.1.2. *Personal firewall*

Personal firewalls have seen an important development related to the deployment of permanent personal access to the Internet via ADSL and cable facilities. However, their use extends outside this framework, because they are also of interest to organizations for protection of laptops issued to employees and used at home, or for teleworking staff.

There are two principal types of personal firewall:

– software installed on a computer;

– external hardware that could be either a device dedicated to filtering functions (sold by firewall vendors, or open source); or equipment incorporating security features (router with a firewall, ADSL devices with firewall functions).

In the case of software that is installed on a computer, personal firewalls are often a product packaged with an antivirus solution. The latter have gradually expanded their product range, offering network filtering and application control functionalities incorporated with traditional antivirus solutions. Firewall vendors also attempt to encroach on this market by offering personal firewall products in software form that can be managed through centralized administration consoles. Security policies are then updated when the person connects to their organization's network.

These policies generally take into account the context of the connection (inside the organization's network, established from a public place, from the user's home, etc.) to adapt filtering rules accordingly.

It is recommended to prohibit the user from changing these settings in case they unintentionally introduce a security vulnerability. It is also important to ensure proper validation/testing of the products and the rules they employ, as any malfunction may result in a rejection of security measures by nomadic employees.

As we have seen, there are also hardware solutions for personal firewalls. These solutions are obviously less mobile than software solutions and apply mainly to teleworkers. In the case of dedicated devices, these should be managed directly by the implementation teams in order to meet the remote access security policies defined within your business. Examples of this type of personal firewall can be found both as commercial solutions and open source products. Apart from dedicated equipment configured by the organization, telecom operators can integrate personal firewalls into their ADSL provisions. However, it is important to be aware that the filtering policies of these ADSL boxes sometimes offer only minimal protection and may not meet your safety standards in terms of filtering.

When you choose your personal firewall, it is critical that you are certain of its ease of implementation (configuration wizards, detection of installed applications, ergonomic configuration GUI, etc.) and use (generation of synthetic reports, filtering of alert messages so as not to be overwhelmed by minor alerts, automatic updates, centralized management tool for remote installation, etc.). A product that is not suitable for your needs or your level of computer skill will give you a false sense of security.

3.3.1.3. *Application firewalls*

With the sophistication of firewalls operating solely at the network layer, attackers quickly realized that their attacks would be easier to carry out (and more successful from their point of view) if they concentrated on:

– the services authorized by firewalls;

– content handled by the application layer (layer 7 of the OSI model), which was not inspected by firewalls focusing only the network portion (layers 3 and 4 of the OSI model).

Faced with these new attacks, "network" firewalls demonstrated their limitations. This shortcoming became even more critical with the publication of the 2006 study by Gartner analysts, showing that 70% of security risks faced by a company are related to the application layer. In this context, application firewalls appeared in the early 2000s. Currently this type of product is mainly concerned with the protection of Webstreams, as well as databases. Two main points should be taken into account in envisaging such a deployment:

– these firewalls function in a "whitelist" or "blacklist" mode. In the first case, the user specifies very precisely what is authorized at the application level. In the second case, the user specifies what is forbidden. Thus, in the case of a whitelist, for access to an organization's Web server, the following will be specified in the application firewall:

- authorized URLs;

- specified variables (names, types, length);

- methods used.

It is therefore clear that in this type of approach, protection can be very fine-grained and effective, but requires good coordination between development services (which modify the code of your Website) and security teams (who manage the application firewall) so that code changes are properly reported to the production level, tested, and validated, particularly in terms respecting security principles;

– the implementation of application firewalls generally exceeds the expertise of the security teams of the organizations which use them. To allow such equipment to be used effectively, it is imperative that the implementation, development and database administration teams work alongside security operators to refine application filtering rules. For example, the structure of a database is only known by the database administrator, and only that person is able to determine which events are unusual or suspicious, and which should not be accepted as requests. But the administrator is not generally aware of the possibilities of an attack on a database, and as such collaborative work between different teams is necessary for successful implementation of this type of equipment.

3.3.2. *Reverse proxy*

The function of the reverse proxy is to ensure a division between a Web resource provided by an organization and the customers who wish to access it. It can thus relay all Web requests to internal server(s) and prevent these servers being compromised. Usually, a reverse proxy is used to relay customers' requests from the Internet to the Web resources of the company. But it can also be deployed in the corporate network to prevent direct access to certain internal networks and specific Web resources.

In recent years, reverse proxy functions have become more complex. At present, these devices integrate modules for application firewalls, SSL accelerators, load sharing (load balancer) and can implement high availability solutions.

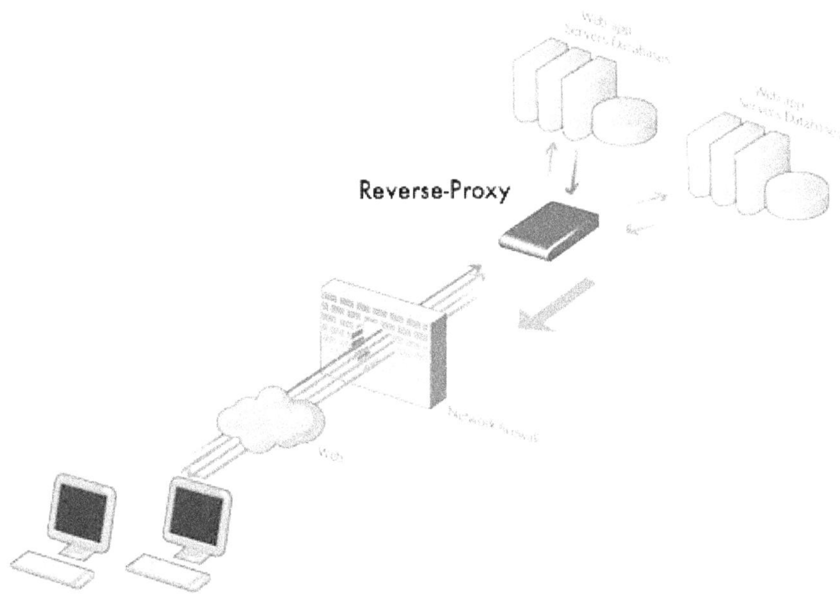

Figure 3.14. *Typical scheme of a reverse proxy architecture*

The reverse proxy is usually located behind a firewall device in the DMZ (DeMilitarized Zone), so that any compromise does not entail the compromise of the resource it is supposed to protect.

The firewall-reverse proxy coupling is generally inseparable, when considering the provision of internal resources from networks whose security cannot be ensured (the Internet).

The current tendency (as with firewalls) is also to deploy the reverse proxy in the form of an appliance (a device whose OS is optimized by the manufacturer for the reverse proxy, and by which administration is simplified, particularly in terms of software updates).

There are two types of reason for deploying a reverse proxy:

– Performance:

1. The reverse proxy can act as an SSL hub, offloading Web server(s) of encryption calculations. The required power is embedded within the reverse proxy and serves all servers.

2. The reverse proxy, due to its function as a load balancer, can balance the load of incoming connections to different servers.

3. The reverse proxy can store the static content of pages provided to customers in a memory buffer, which increases performance.

– Security:

1. Customers do not have direct access to the fundamental resource, because the reverse proxy adds an additional layer of protection that makes attacks more complicated.

2. Accessed resources are isolated, because the reverse proxy is not in the same domain as the destination servers.

3. The reverse proxy can use many filtering rules (access methods, site names, resource requirements, etc.) and can even perform complex application control (white list of permitted parameters, length of variables posted, type of variables, etc.) if the firewall application modules, that certain products offer, are activated.

3.3.3. *Antivirus software*

Antivirus software is a major component in the battle against viruses. However, as we will see in section 3.4 – *Organizing the information system's security device*, antivirus software alone is not sufficient, and virus security should be thought of from a broader perspective, lest it is easily defeated.

Both functions are assigned to antivirus software: first, detection and analysis, and second, repair of the infected file when possible.

3.3.3.1. *The various detection techniques*

There are of two types of virus detection technique: static and dynamic.

Static techniques

These are based on:

– the use of signature files (or pattern matching) using a database in which each virus has a very specific identifying signature, in the same way as your fingerprint identifies you individually, such that you cannot be confused with another individual. This approach has the merit of working well for known viruses, but also has two major drawbacks:

1) there is no guarantee that a mutated virus whose code has been modified can be identified by the signature of the mainstem;

2) the creators of viruses study the way in which existing antivirus software functions to find workarounds so that their creations are not detected by these signature-based approaches (encryption of the virus code, etc.);

– the use of heuristic methods: we seek here to detect abnormal behavior in the analyzed programs on the basis of rules, strategies, etc., to determine whether that behavior is caused by a virus or not. The main difficulty with this approach is that these methods can generate false positives, and programmers of viral codes can also analyze these rules to find out how to circumvent them.

Dynamic techniques

These techniques are based on the study of what goes on at the level of the machine, via behavioral analysis of programs, or by using a sandbox, where the software is tested in a virtual environment to establish whether its actions are malicious.

The main drawback of this approach is that it consumes significant computing power, and is penalizing in terms of performance, since the antivirus scans everything that occurs. That is why it is usually reserved for on-demand analyses.

3.3.3.2. *How to choose your antivirus software*

The criteria presented in Figure 3.1 and detailed in Table 3.4 should be taken into consideration in choosing an antivirus product in the context of your business activities.

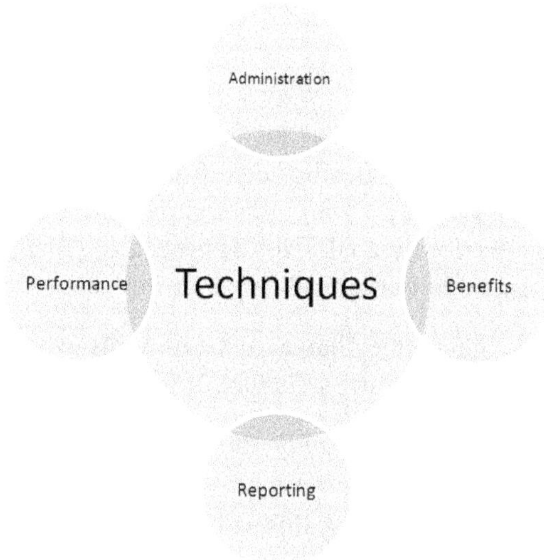

Figure 3.15. *Criteria for choice of an antivirus solution*

Criteria	Points to take into account
Techniques	– The methods of detection used (signature files, heuristics, behavioral, sandboxing, etc.). – Limitations on the types of file which can be processed. – Compressed file types which can be processed, and their parameters (total number of directories, subdirectories, number of files contained in the compressed file, etc.). – Types of stream which can be scanned (Web, email, etc.). – The possibility of forbidding the disabling of software protection by the workstation user.
Administration	– The ease of use of the administration console. – The ease of configuration of the antivirus parameters (exclusion of directories, exception possibilities, scan timetabling). – The control functions of the console, to verify the antiviral coverage of workstations (some products offer the ability to verify whether the antivirus is active on certain workstations by scanning entire network). – The availability of the product on different OS (Linux, Windows, Mac OS X, etc.).

Table 3.4. *Principal points to take into account in choosing antivirus software*

Reporting	– The possibility of producing reports adapted to the populations to which they are destined. – The richness and relevance of the information reported. – The sending of reports via email at regular intervals.
Performance	– The possibility of setting the maximum workstation's CPU and memory use by the antivirus (antivirus which slows down workstations penalizes business activity).
Benefits	– The list of certifications (ICSA, West Coast Labs, Checkmark, etc.) and recognitions (VB100 award, SCMagazine Award, PCUser) obtained. – Results obtained in the IAWACS challenges.

Table 3.4. (continued) *Principal points to take into account in choosing antivirus software*

NOTE.– For SMEs and self-employed individuals, there are very flexible and easy to use solutions, based on the use of Web servers, which involve downloading an ActiveX or Java applet which can then scan the contents of the hard disk of the user's computer for viruses.

3.3.4. *Antivirus software: an essential building block but in need of completion*

The current situation does not allow us to consider that the establishment of an antivirus system will be sufficient to guard against future infections.

Good antivirus protection can and should be organized as part of a more global reflection of your business risks. More specifically in order for the antivirus system to provide relevant and efficient protection, it must be supplemented by the following:

– monitoring of applications installed on workstations. If installation of a program is not carried out by an authorized person (system administrator, etc.) it must be prohibited by adequately configuring the operating system or employing additional software. Controlling the execution of programs that are not part of a whitelist is also good practice;

– integration of antivirus protection in the security device by:

- putting in place detection probes (NIPS, NIDS) at sensitive points in your organization's network to detect viral behavior, worms, trojans, etc.,

- taking into account *firewall* configuration by blocking traffic that may be specific to the virus (e.g. Web server belonging to a hacker that allows the virus to update in order to circumvent the antivirus software),

- the creation of a system for alert correlation or an SIEM;

– raising your users' awareness of their responsibilities and duties with respect to the information system. Education of staff on security topics in order to obtain their involvement in and observation of security policies is essential to prevent and reduce virus-related incidents.

3.4. Organizing the information system's security

We have often made parallels with military methods in their approaches/methods of attack or defense and the protection of information systems. But until now we have not discussed the organizational part of a defense mechanism. However, when discussing the military, no-one disputes the need for an army that is sufficiently trained, disciplined and well-organized to carry out its missions. After all, the very essence of an army includes:

– knowing the potential threats it may face;

– knowing how to react;

– knowing how to mobilize resources to meet attackers;

– establishing a chain of command in order to know how to react to crises, attacks, and lead the troops in carrying out their tasks.

In civil matters, concerning the system of information security, although safety equipment and training of operating teams is often a focus, the organizational aspect is sometimes (often!) neglected or postponed. Yet this is the necessary cornerstone for the proper functioning of the whole. As we have mentioned, who could imagine that an army has a weak chain of command, poorly trained soldiers, not knowing how to recognize their enemies, nor the threats they must face? However, frequently outside the military, it is clear that the organizational aspect is often the most difficult to implement or maintain. When one thinks of organization of security, the terms training, processes, procedures, standards, etc., come to mind, usually with negative connotations:

– awareness:“it takes time”, “it’s tedious”, etc.;

– procedures: “they’re constraining”;

– standards: “they’re not adapted to the context of this business”.

This situation stems mainly from the fact that security is a domain which brings together technical experts and other specialists in business standards and processes. These two worlds nonetheless share a common goal: to preserve the security of the information system.

3.4.1. *What is security organization?*

Certain accepted wisdom summarizes or reduces the security organization of a information system to the regular raising of awareness with staff, the dissemination or regular recall of internal rules, policies related to Internet or messaging use, and compliance with various procedures (policies regarding flow opening, securing new servers, information classification, etc.).

Security is therefore perceived as a more or less coherent set of building blocks that constrains users’ experiences, because they are not aware of the end purpose.

Evidently, the organization of security is the antithesis of this received wisdom. Organizing the security of an information system is principally:

– to define objectives in terms of information security (threat assessment, etc.);

– and once these objectives are achieved, to ensure that this situation will continue (regular audit).

If you follow these principles, you will have established a clear command structure, you will have evaluated your threats and taken the necessary decisions to respond to them, with trained personnel in a regularly re-evaluated framework. This is what is known as an ISMS (information safety management system).

3.4.2. *Quality of security, or the attraction of ISMS*

An ISMS is a particularly effective tool for ensuring that all your security, in the broadest sense of the term (equipment, staff, organization, etc.) is adapted to cope with the threats that you may encounter, bearing in mind the objectives that have been set in terms of IS security.

To define an ISMS, two aspects are fundamental: a standard and a management system.

A standard is the intersection of the three criteria shown in Figure 3.16. A management system is the modelization depicted in Figure 3.17.

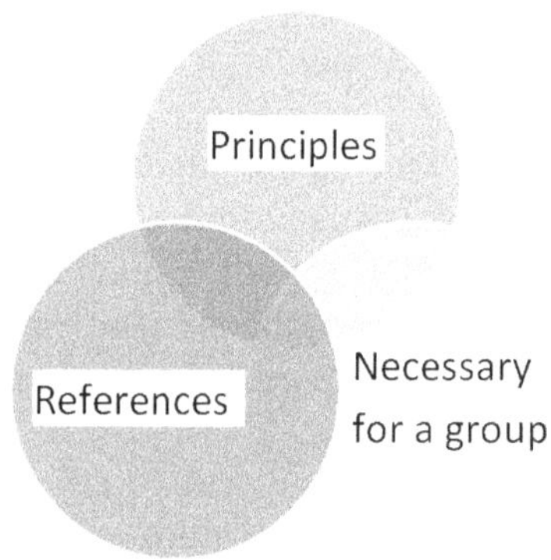

Figure 3.16. *Constituents of a standard*

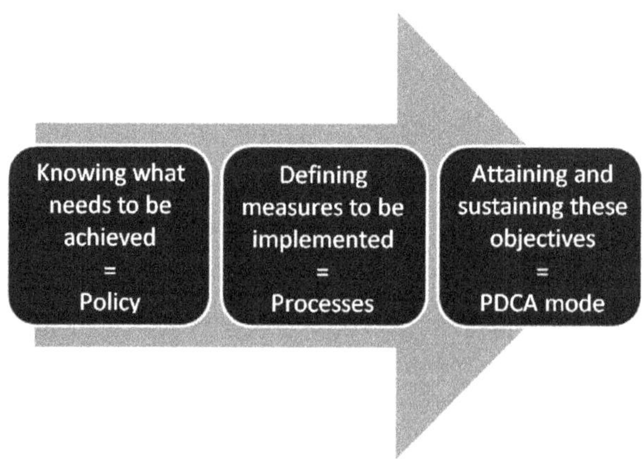

Figure 3.17. *Schematic of a management system*

The PDCA (Plan Do Check Act) mode refers to a method of quality management, which is conducted according to four sequential phases:

– *Plan*: define what you wish to achieve;

– *Do*: implement what has been defined at the planning stage;

– *Check*: check that everything is functioning as anticipated with respect to the decisions implemented in the previous stage;

– *Act*: correct anomalies detected at the previous stage, and if this leads to new tasks/decisions to be realized, we loop back to the start and the cycle begins again.

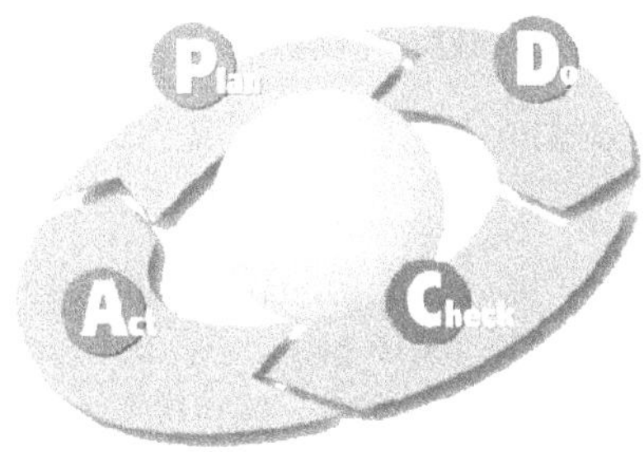

Figure 3.18. *PDCA model of continuous improvement*

The standard which refers to the security of information systems is ISO 27001. This text, the cornerstone of the organization of your security, is essential if you wish to structure your defense system effectively and sustainably.

There are two main points to remember on this subject: its implementation is based on the construction of nine organizational processes that we will not describe in detail here, but naming them is sufficient for discussion with respect to their content. This standard, by implementing these nine organizational processes,will allow you to:

– build (organize) the lines of defense of your information system, with respect to your security objectives;

– be consistent between the defenses implemented and risks which can threaten your information system, and thus avoid unnecessary safety equipment and/or procedures.

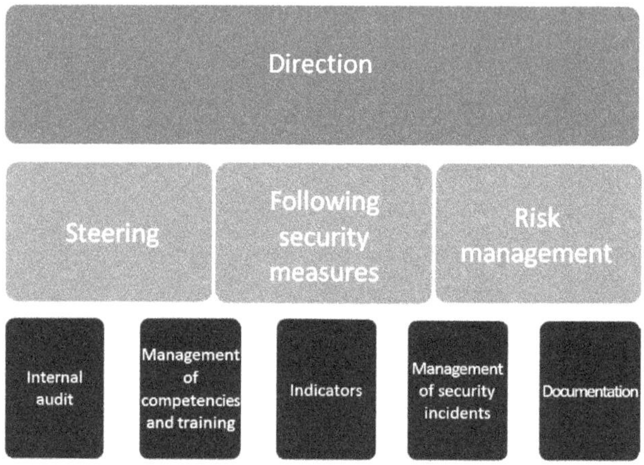

Figure 3.19. *Structure of the ISO 27001 standard*

The most frequently heard criticism with respect to the ISO 27001 standard, or against an ISMS more generally, concerns the complexity of the implementation and therefore the costs incurred.

However, this security structure is never questioned when it comes to military matters, in terms of discipline or the chain of command, and which itself contributes to the effectiveness of an army.

Chapter 4

Technological Countermeasures for Remote Access

"The depth of the walls is less important than the will to defend them"

Thucydide

Connecting a mobile device to organizations' resources must at the same time meet the expectations of the user (who wants working conditions similar to those in their office) and the security constraints of technical teams, who want above all to ensure the integrity of the information system. Achieving these seemingly contradictory needs has required:

– the development of specific tools to allow remote connections, or the adaptation of existing solutions;

– the implementation of specific access controls which are more stringent than those generally in place within the organization's LAN;

– the definition of architectures adapted to provide the services required while addressing the risks specific to mobile access.

All of the above will be discussed in this chapter, starting with the basis of all mobility tools: remote connection solutions.

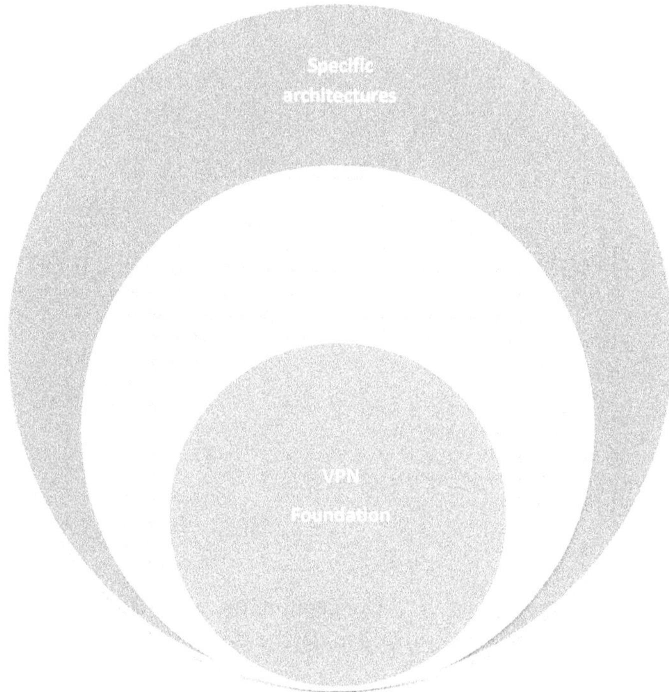

Figure 4.1. *General architecture of remote access solutions*

4.1. Remote connection solutions

There is a multitude of technical solutions for enabling mobile devices to connect to organizations' IT resources. These solutions can be managed directly by the company or subscribed to through services offered by a provider (telecom operator, for example).

The implementation of such a solution is not confined simply to installing a "black box" for establishing connections. It must also be linked to systems for authentication and filtering access.

This implies that associations must be established between levels of resource sensitivity and levels of user authorization prior to accessing the system. The channel by which an individual tries to access the resource (VPN network via the Internet, telephone access, etc.) can directly affect the authorization decision (e.g. denying remote access to the salary database).

4.1.1. *Historic solutions*

The need to be able to remotely access and share IT resources did not suddenly arise in the 2000s – it has always existed. It was to satisfy this need, amongst others, that researchers developed the Arpanet network in the 1970s.

They put tools in place which subsequently became indispensable, such as telnet and RLOGIN, for connecting from a remote computer, or RCP and FTP for transferring files.

At the time, security requirements were not as imperative as they are today, as these tools were accessible only to a small number of people who belonged to the same community. Thus, for example, authentication mechanisms as they then existed were no more than plain text exchanges (unencrypted data) of a username and a password via the network.

Of course, with the popularization of access to these systems, hackers did not take long to discover all aspects of security vulnerabilities of which they could take advantage.

Developers have therefore evolved these tools so that they now integrate security mechanisms such as strong authentication, data exchange encryption, control session integrity, etc., in order to reduce the risks associated with their use. This includes, for example, the secure versions of RCP (→ SCP), RLOGIN (→ SLOGIN) and FTP (→ SFTP) that have been developed by the company Sun for its Unix Solaris system.

But without doubt, the program most adopted in the Unix world is Secure Shell (SSH). It allows, like RSH, connection to a remote computer, but unlike this program, it ensures the integrity and confidentiality of data exchange (RFC 4253) and allows many methods of user authentication (RFC 4252). Furthermore, it allows the creation of a tunnel between two computers and therefore the securitization of applications/protocols that are not natively protected.

4.1.2. *Desktop sharing solutions*

Whether the *Remote Desktop Feature* included in the professional version of Windows, or the extended functionality of collaborative tools or specialized software such as Symantec's *pcAnywhere* or Citrix's *GoToMyPC*, there are numerous solutions for remotely controlling a computer via a PSTN network or the Internet. Using these solutions the user can interact remotely with the computer (using their own keyboard and mouse) of which they have taken control, and see an exact copy of the remote display on their screen. They therefore have access to the same resources, applications, etc., as if they were sitting directly in front of their office computer.

While initially this type of solution was developed so that the technical assistance services could remotely "hold one's hand" to address the various problems that the end user might encounter, or so that the system administrator could perform maintenance operations from home, many people today use this type of tool to access business resources via their office computer.

If on a practical level this solution seems perfect, it could even be used from a Web browser, it nevertheless poses numerous issues of security. Confidentiality of data exchanges that are not always encrypted – or are encrypted using proprietary algorithms whose robustness has not been proven – is but one. Furthermore, when such exchanges are encrypted, protection software installed on the network (*firewall*, IPS, etc.) is rendered "blind", and therefore cannot ensure that these flows contain no risks.

But most importantly, this type of solution gives rise to an increase in the number of computers that can serve as bases from which a hacker can launch an attack. Since these computers are largely desktop PCs without enhanced security protection, they can be compromised much more easily.

NOTE.– It is not necessary to have a PC to use this type of solution. Many vendors, beyond offering a solution based on the use of a Web browser, have also developed specific clients which can be installed on smartphones based on Windows Mobile.

4.1.3. *Publication on the Internet*

With the spread of Internet connections, one of the simplest solutions for providing remote users with access to a limited number of business applications is to "publish" them on the Web.

If the resources in question are native Web applications, this will not be a particular problem. If not, most client-server software vendors now offer solutions for adapting their content to the Web.

For example, electronic mail – which has become the flagship application for most businesses – is usually offered with a Web interface, commonly referred to as *Webmail*. This enables people on the move or at home to connect via any browser and check their work email.

There is also the possibility of using a remote desktop solution such as Microsoft Terminal Server or Citrix tools. This type of solution has many advantages, for example:

– not having to oversee installation of a proprietary software client on all workstations;

– economizing on license fees related to the installation of the software client on numerous workstations;

– restricting access to just the application that is "published" on the remote desktop;

– providing the same level of functionality and the same workspace experience as the proprietary software client, where the "webified" solution may perform less well in these respects.

Of course, it is best to place such servers in the DMZ so that they are protected by the *firewall*, or to use a proxy server installed in the DMZ to relay requests to the Intranet

It is also possible to augment the security of this infrastructure with an *application firewall* to thwart attempted Level 7 attacks.

NOTE.– A variation of this principle consists of establishing a portal (Website, terminal server, etc.) whose content may or may not be personalized according to the user, and which provides access to multiple applications. In this way, an organization only needs to manage a single point of entry between the exterior domain and the resources it wishes to make available.

4.1.4. *Virtual Private Network (VPN) solutions*

Almost all current networks in existence today use IP V4. This protocol, of which the first version was created in the late 1960s, was not designed to take into account security issues; its primary purpose was to enable resource sharing and dissemination of information, rather than protection. It is therefore relatively easy to intercept or alter data transmitted using this protocol[1].

When this protocol emerged from the university where it was created and began to be used by companies, the question of security arose.

Alongside grew the need for companies to be able to extend their corporate LAN to very remote physical locations, and preferably through a shared network infrastructure such as the Internet, in order to lower the cost.

It was to address these two requirements that virtual private network (VPN) solutions were developed. These VPNs, as we shall see in more detail in the following sections, incorporate cryptographic mechanisms that encrypt data and verify their integrity and authenticity. This ensures that the information exchanged is not read or modified by a third party, and that the partners with whom you communicate are indeed who they claim to be.

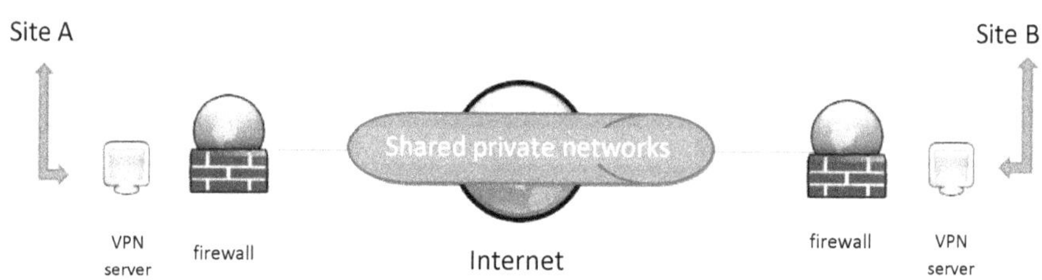

Figure 4.2. *Example VPN structure*

A VPN consists of creating a virtual "tunnel" through a shared infrastructure[2] in order to transmit data from site A to site B. To create this

1 The *wireshark* tool, for example, allows access to all network traffic circulating on a LAN to which your machine is connected.

2 The term shared infrastructure is used when data are routed through a shared (or public) network such as the Internet, and private infrastructure when using a dedicated service marketed by a telecoms operator.

tunnel, previously encrypted data are encapsulated in an IP V4 protocol layer. Data encryption is performed automatically at the tunnel entrance, and data decryption is performed at the exit.

The solution is known as *private* because only the A and B sites are aware of the data exchanged through the tunnel. It is known as *virtual* because it is a logical tunnel established over a shared physical infrastructure.

VPN solutions have experienced strong growth in recent years because of their low cost and the ease of connection to an organization's information system they offer to:

– mobile employees;

– small businesses;

– partners (clients or providers).

In addition, the deployment of VPN solutions has been facilitated by the integration of this feature as standard in most security equipment (*firewalls*, routers) and operating systems.

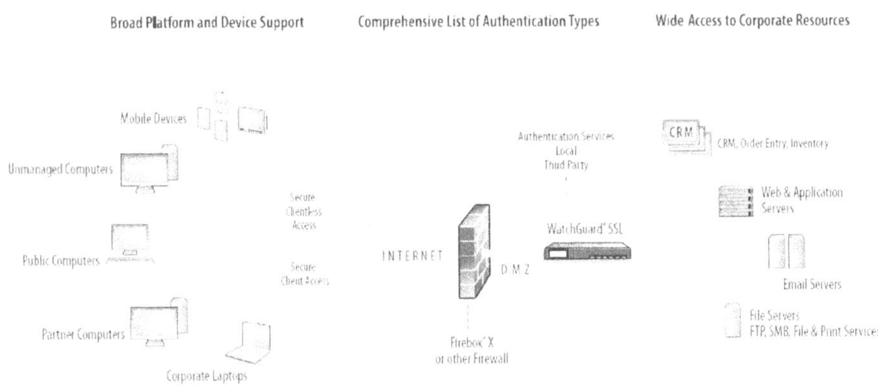

Figure 4.3. *WatchGuard SSL 100 Solution © WatchGuard*

4.1.4.1. *Different VPN technologies*

Several protocols allow a VPN to be constructed, but they do not all offer the same type of services. The specific characteristics of each must therefore be fully understood in order to choose the solution that best suits your needs and constraints.

4.1.4.1.1. Layer 2[3] tunnel – PPTP, L2TP

A layer 2 tunnel will carry frames between two points of interconnection, regardless of the network protocol (layer 3 of the OSI model) used (IPX, NetBIOS etc.). The advantage of being independent of a network-type protocol is now relatively small, because IP has supplanted all other protocols. These are now used only on rare occasions (obsolete equipment for which there is no IP layer, etc.).

Currently, two main families of layer 2 tunnel exist: PPTP and L2TP. PPTP was developed by Ascend, Microsoft, 3Com and U.S. Robotics. L2TP is the result of the merger between Cisco's L2F (Layer Two Forwarding) and PPTP. It was then standardized (RFC3193), under the auspices of the IETF.

PPTP is very popular and can be found on most equipment, as it has been included in all Windows operating systems since their version 95. The security features offered by L2TP are better, however, as it includes two functions that do not exist in PPTP: data integrity management and non-repudiation.

4.1.4.1.1.1. PPTP (Point-to-Point Tunneling Protocol)

PPTP is an extension of the serial link IP packet transfer protocol, created in 1994 by the IETF (RFC 1661) and known as PPP, of which the authentication, compression and encryption functions have been strengthened.

To this end, the PPP frame is encrypted using MPPE (Microsoft *Point-to-Point Encryption*) with a key generated using MS-CHAP, MS-CHAP v2 or EAP-TLS. The frame is then encapsulated in a GRE (*Generic Routing Encapsulation*) packet, which indicates the client workstation address and that of the VPN concentrator in the IP header, so that it can be sent over the network.

---

3 ISO defined a seven-layer model for designing communication protocols. Layer 2, referred to in the title of this section, corresponds to the "link" layer of this model.

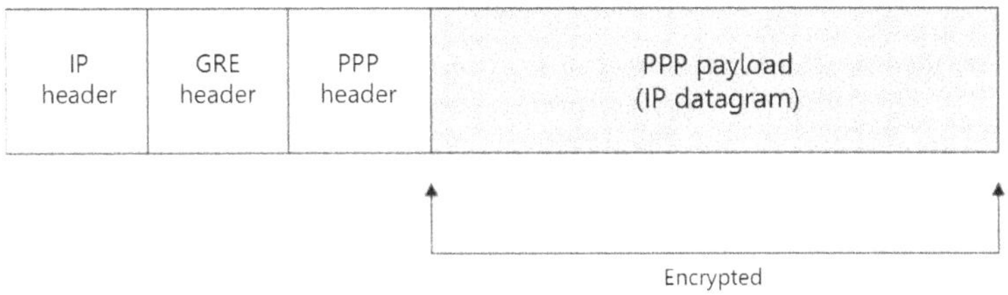

Figure 4.4. *PPTP frame*

It should nonetheless be noted that although PPTP supports data confidentiality (encryption) it does not guarantee integrity or non-repudiation.

4.1.4.1.1.2. L2TP (Layer Two Tunneling Protocol)

The PPP frame is first encapsulated in an L2TP packet before being encrypted using "IPSEC encryption" mechanisms. Then the L2TP packet is encapsulated in an IPSEC packet that contains an *Encapsulating Security Payload-Authentication Trailer* for ensuring integrity and message authentication. It also contains an IP header which indicates the address of the client and the VPN concentrator so that the packet can be routed over the network. L2TP therefore natively provides a layer allowing tunneling, and IPSEC (discussed in the next section) provides the security features.

L2TP can use PPP authentication methods, and also other more advanced systems such as RADIUS and TACACS+.

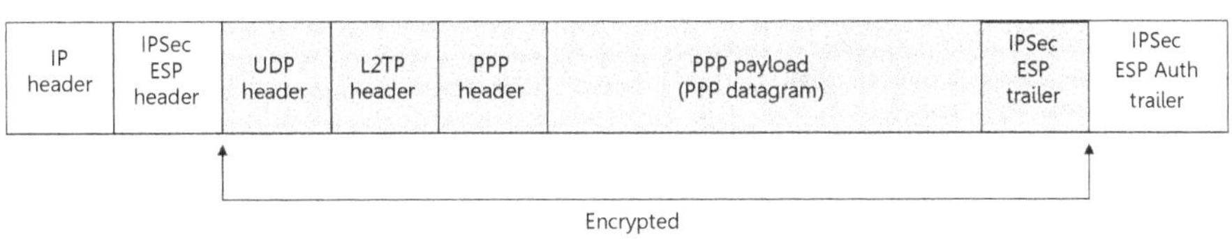

Figure 4.5. *L2TP frame*

4.1.4.1.2. Layer 3 IPSec tunnel

IPSec (IP *Security*) is a set of IETF standards originally developed for IP V6. However, as this new version of the IP protocol has not yet been widely deployed, IPSec has been adapted for the current version of the IP protocol

(IP V4) in order to add the identification, authentication and encryption functions it previously lacked.

It can be used in two modes:

– *transport mode*: only the "payload" of the IP packet is "protected" [4]. It is used for *end-to-end*-type communications, such as those that take place between a computer and a server. The name stems from the fact that IPSec can protect the transport layer of the OSI model, although in reality its scope is broader and can, for example, be used with ICMP;

– *tunnel mode*: the IP packet is "protected" in its entirety, and a new IP header is added to it so that it can be routed. It is used for communication between two computers, between a computer and a network or between two networks.

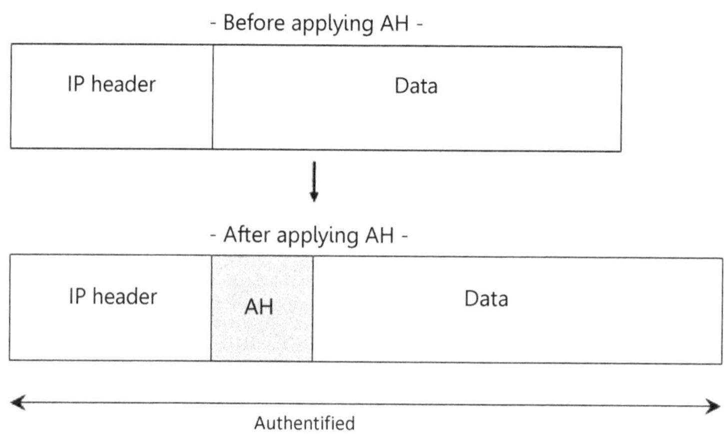

Figure 4.6. *AH in transport mode*

To do this, IPSec relies on several mechanisms that can be used separately or conjointly.

– **AH** (Authentication Header) (RFC 2402)

This ensures authentication of the sender of the packet as well as the integrity of its contents (data + IP header). To achieve this, AH uses a

4 Protection modes are explained in the next section.

hashing algorithm (HMAC-MD5, HMAC-SHA1, AES-XCBC-MAC, etc.)[5] that calculates a digital "signature" from the data in the datagram and a secret key shared by the sender and the receiver.

Of course, in the transport mode only the fields that are not variables in the IP header will be subject to *hash code*[6] calculation. If, for example, TTL were incorporated, it would modify the value of the signature with each pass through a router.

AH also allows the non-repudiation of the packet to be ensured, since the sender can neither deny having been its originator, nor its uniqueness. These various pieces of information are stored in the "AH *header*" that is inserted into the IP packet between the original IP header and the "payload" in transport mode, and before the header in tunnel mode.

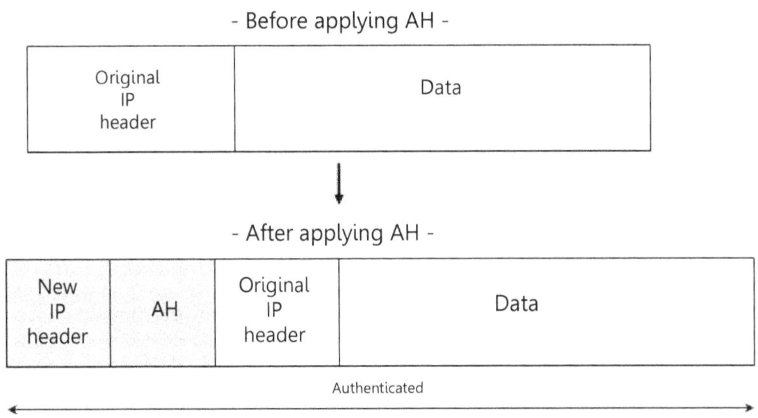

Figure 4.7. *AH in Tunnel mode*

– **ESP** (Encapsulating Security Payload) (RFC 2406)

Like AH, ESP provides authentication, data integrity, and protection against *replay*-type attacks. In addition, however, it ensures confidentiality of both data (tunnel/transport mode) and the IP packet header (tunnel mode)

5 In RFC2402, the choice of hashing algorithm is not imposed, only support of HMAC-MD5-96 and HMAC-SHA1-96 is mandatory.

6 Fields not taken into account in the calculation of the "fingerprint" are: *type of service*, *fragment offset*, time to live, IP *header checksum*.

by using a symmetric encryption algorithm such as DES, 3DES AES-CBC[7], AES-CTR[8]. It should nonetheless be noted that integrity control is less comprehensive than with AH, since it does not cover the entire IP packet.

Information is stored in two additional fields (ESP *header* and ESP *trailer*). The first, the ESP *header*, is inserted between the original IP packet header and the "payload" in transport mode, and before the header in tunnel mode. The second, the ESP *trailer*, is inserted at the end of the IP packet whether in transport mode or in tunnel mode.

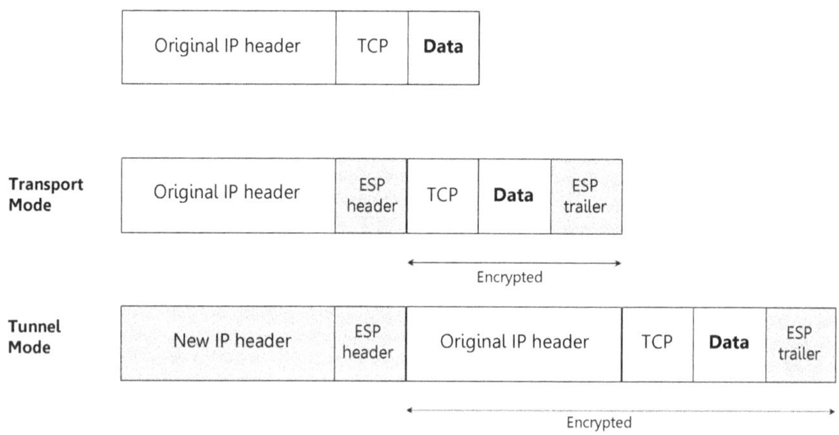

Figure 4.8. *The different modes of Encapsulating Security Payload*

– **IKE** (Internet Key Exchange) (RFC 2409)

IKE allows two communication partners to negotiate the choice of cryptographic algorithm (AES, 3DES, etc.), the method of integrity control (HMAC-MD5, HMAC-SHA-1, etc.) and the authentication mechanism (*pre-shared keys*[9], digital signatures, etc.).

It also ensures the exchange of encryption keys as well as the mutual authentication of the communication partners. It is a hybrid protocol, based on:

7 AES *Cipher Block Chaining*.

8 AES *CounTeR Mode*.

9 Encryption keys which have been shared between two partners before the establishment of a communication in order to allow its establishment.

– Oakley (RFC 2412), from which it takes the concept of using different modes for key exchange;

– SKEME, from which it has taken the mode of operation of shared key authentication and fast key refresh;

– *Internet Security Association and Key Management Protocol* (RFC 2408), from which it takes key exchange and the establishment of SA (*Security Association*).

– IPCOMP (IP Payload COMpression Protocol) (RFC 3173)

IPCOMP is used to compress the data carried in IP packets before they are encrypted by ESP, thereby reducing bandwidth use.

Because it is a low-level protocol (in contrast to SSL, which we discuss in the next section), it secures all types of exchange. This solution is therefore widely adopted for building tunnels between two sites. Thus, when the tunnel is opened, the client or site connected through it to an organization's network has full connectivity, as if it were connected to an extended LAN.

This is also one of the main criticisms leveled at this solution in terms of safety, because once the user is logged in, they have access to all the company's network resources. A solution exists – specifically, by using IP/port filtering (ACL) – but this technique is relatively cumbersome to implement, and especially to manage over time (changes of servers' IP addresses to manually take it into account). In addition, event logging functions are of low granularity relative to the wealth of information which can be recorded, which restricts the range of possible analyses during a security or operational incident.

The other main disadvantage is that making this connection requires an IPSec client which must be specifically configured. For this reason, this solution is relatively complex to deploy on a large-scale, across numerous workstations (IPSec client installation, configuration).

In addition, because the IPSec client must first be installed on a computer before this solution can be used, it is not very well adapted for mobile users who need access to business resources at all times and from any type of terminal (free PC service at an airport, home computer, smartphone).

NOTE.– In order to ensure that the PC which establishes the tunnel does not serve as a bridge between the organization's network and a possible hacker on the Internet, it is important to disable the split tunneling option in the configuration of the VPN client. In effect, this feature allows a user to send IP packets to the servers of the company through the tunnel and other data (surfing Websites) directly via the network to which he is connected (Internet), whereas without this function all flows through the IP tunnel. For additional security, this option must be automatically disabled by the remote access server when establishing the tunnel, rather than it being left to the discretion of the user.

For additional security, this option must be automatically disabled by the remote access serve when establishing the tunnel, rather than it being left to the discretion of the user.

This also ensures that only the computer's network interface used to establish the IP tunnel will be enabled and that the other network interfaces (Wi-Fi, second Ethernet card, MODEM card) will be disabled automatically.

NOTE.– For the sake of scalability, the designers of IPSec did not specify any particular signing algorithm (except for MD5 and SHA-1) and encryption (except DES). In addition, certain vendors do not implement the standard in the same way (for example, when the new SA should be used) or have added features which were not, or poorly, covered. It is therefore sometimes difficult to communicate with IPSec solutions from different providers.

Because of this, the *Virtual Private Network Consortium* conducts interoperability testing between different solutions on the market, the result of which can be viewed on their Webserver (www.vpnc.org/testing.html) and even delivers certifications. You can also refer to the ICSA lab Website (www.icsalabs.com/IPsec) which has developed a protocol for verifying the compatibility of different implementations of IPSec.

NOTE.– Using the IP address translation functionality (NAT) can cause problems when used in conjunction with IPSec (AH or ESP function in transport mode) because, as we have seen, this protocol verifies the integrity of its data, and a change of address in an IP packet is considered a change of the same type as an attack by a hacker.

Nonetheless, several methods of solving this problem exist, such as using the NAT function in the same equipment that manages the IPSec tunnel, or the use of NAT Traversal (NAT-T). This is based on the encapsulation of IPSec packets in UDP. Thus, it is UDP and not IPSec packets which are modified during passage through the equipment providing network address translation (NAT).

NOTE.– It is also important not to forget to configure the remote access server to automatically terminate any connection after a certain period of inactivity. This is to ensure that a computer left unattended, on which the screensaver is not activated, will not be used by an unauthorized person.

For this feature to be truly effective, it is essential that the client or remote access server analyzes the activity of the workstation so as ignore automatically generated traffic, which may be the case with messaging software which consults the inbox at regular intervals.

To avoid keeping unnecessary information on the IPSec connection (SA), an additional protocol called *Dead Peer Detection* is used, which queries the remote device at regular intervals to ensure it is still connected. If the remote device does not respond to this request after the *timeout* and the configured number of repetitions, then all connection information is deleted.

NOTE.– Some implementations of the IPSec client store the *pre-shared key* on the hard drive without encrypting it beforehand. Thus, a person with access to the computer or software that managed to access it, can easily read it.

4.1.4.1.3. Layer 5 tunnel - VPN SSL (*Secure Socket Layer*)

SSL VPN solutions have the advantage that they can be used from all types of mobile terminals and can easily be deployed/maintained. They have consequently experienced strong growth in recent years, because they meet the growing needs of business connectivity.

This ease of deployment is due to the fact that these solutions are based on a standard protocol (SSL) which is present natively in all browsers on all devices (PC, smartphone) and is one of the services authorized by almost all *firewalls*. The use of SSL to mount a VPN tunnel can be achieved using any device, regardless of geographical location, as long as it has Internet access. It is not necessary to install special software on the client: a Web browser

(Internet Explorer, Firefox, Safari, Opera) will suffice. There are, therefore, no additional complications introduced in terms of deployment and maintenance.

In addition, SSL VPN solutions provide a high level of security because access restriction can be limited to specific applications/resources and not to the whole network, as is sometimes the case with classical tunnel solutions. The user connects through a Web portal of which the content has been personalized with the permissions allocated to the user by the administrator. Thus, the user can only access the applications/resources that appear on the personalized homepage on the Web portal. There are different types of SSL solution: clientless connection, thin client connection, thick client connection.

Clientless connection

Historically, the first type of SSL VPN solution was to implement a personalized portal to provide a remote user with access to various business applications after that user has been authenticated. The portal plays the role of an HTTP gateway (proxy) between the end user and internal servers, by converting the private IP addresses of an organization's network into shared addresses. The main disadvantage of this architecture is that it can only provide access to Web applications. Nonetheless, this can be more than enough for occasional access (email access) or to give access to some applications to business partners such as suppliers or customers.

Thin client connection

The second generation of SSL VPNs allowed the expansion of access to non-Web applications and potentially all applications and business resources. To do this, the user automatically downloads from the VPN gateway, a small program written usually in Java or ActiveX. This intercepts the messages of non-Web applications and transmits them through the SSL tunnel. The SSL gateway then assumes the role of "translator" of these non-Web applications (SAP, file management system) so that they can be used from the browser.

This extension of SSL functionality results in a corresponding limitation of universality of access, since only platforms capable of supporting the technology (Java, ActiveX, etc.) used for the client software could use it. Thus, client software based on ActiveX will only work on Windows platforms. In addition, the installation of plug-ins is not always allowed on publicly accessible computers.

Thick client connection

As in the case of thin client access, software is usually downloaded from the VPN gateway through the Web browser. But it not only provides an SSL tunnel, it also provides many other extended features. Among them are the following:

– verification of compliance with the client's security policy: the software will verify that the computer has the latest patches and there is no *malware* installed;

– installation of a virtual desktop on the client: the user works in a secure environment that ensures that data are not compromised during the session and are removed from the workstation when the session ends. This solution is particularly suitable where the company cannot guarantee the conformity of the equipment used (self-service PC) – this provides a means to ensure that it will in no way compromise the security of the information system.

Again, as in the case of thin client, the issue of universality of use of this solution arises, but to an even more important degree, because to achieve these advanced features, the software needs certain access privileges on the machine, which will not necessarily be authorized.

SSL solutions using thick or thin clients usually have the Session Cleanup function. With this, the SSL client automatically removes all traces of the connection on the user's computer (cookies, URL history, Web pages placed in the browser cache, downloaded files, entries in the registry, temporary files, etc.) once the session is completed. This feature is even more essential when the user has used a self-service computer (e.g. cybercafe) in order to connect, because if this "housekeeping" is not automatically carried out, a malicious person could attempt to retrieve confidential information after the user had ended their session.

NOTE.– It is important to note that in the context of SSL VPN solutions, only the TLS protocol has been standardized. Thus, all additional functions such as verification of the workstation's compliance to the company's security policy, or the creation/management of the SSL tunnel, are carried out using the proprietary solution. The main consequence of this is that it is not possible to set up and use a heterogeneous solution from multiple vendors.

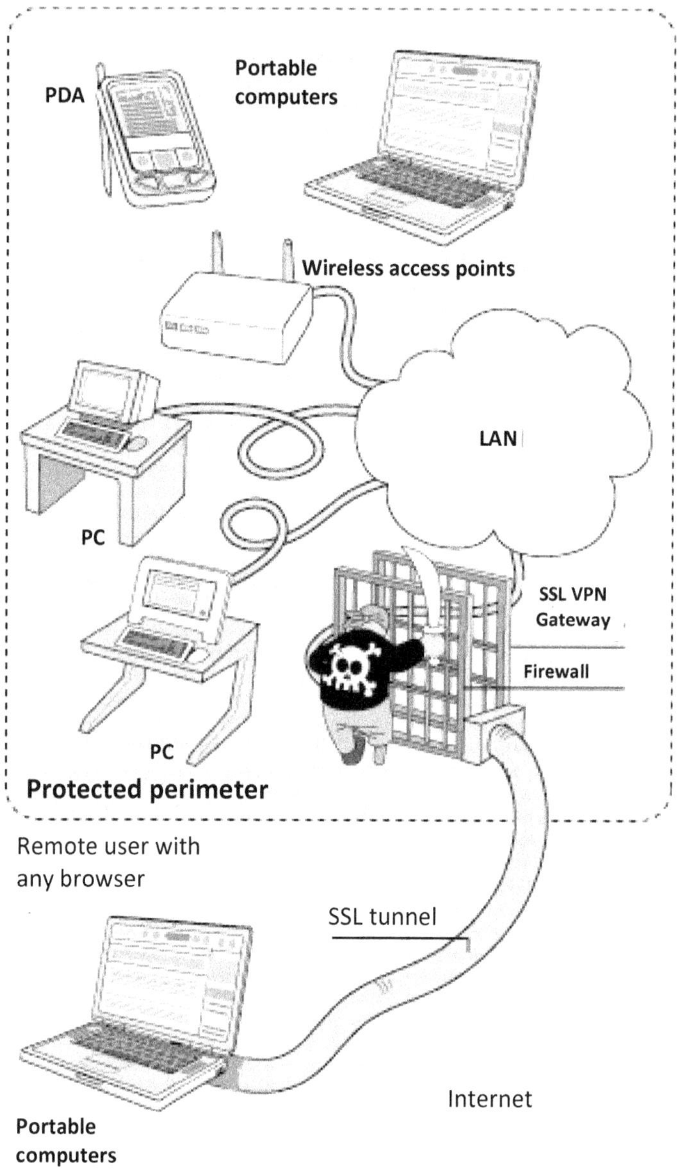

Figure 4.9. *SSL VPN Solution© Walloon Telecommunications Agency (www.awt.be)*

4.1.4.2. *Access policies*

The latest VPN solutions, especially those based on SSL, not only allow us to establish a tunnel between the client and the corporate network, but can also now adapt resources' access permissions based on many parameters such as:

– the identity of the user/predefined group membership (for example, someone belonging to the system administrators group will have different access rights from an employee of the Human Resources Department);

– the type of equipment used to establish the connection (for example, a smartphone that does not have the same level of functionalities as a PC, will have more restricted access than the latter, such as the ability to mount a remote directory);

– the type of operating system used and level of patch applied (for example, an earlier version of Windows may be considered less secure and only have limited access);

– the security level of the client computer (for example, depending on the presence or absence of a certain number of security tools such as antivirus, *firewall*, etc., and their updates, the level access may be changed), and if protection software is active at the time of the connection request;

– authentication mode used (e.g., if the user has employed a disposable password or OTP, they will have a higher level of access than another person who has only used a simple password authentication);

– the type and version of the protocol used to secure exchanges (e.g. TLS 1.0 is considered safer than older versions of SSL);

– the owner of the equipment used to establish the connection[10] (for example, a PC belonging to the organization will be considered safer than a personal computer, and therefore will have a privileged access level);

– the place from which the user establishes a connection[11] (for example, a person working from home may be considered to be in a safer place than if he was in a cybercafe);

---

10 To determine if a device belongs to the organization, it is possible to use a certificate pre-installed on the computer, for example, or to create a dedicated entry in the registry database.

– the date and time (for example, if the person works during non-working days, he may have limited access, as employees responsible for security control have fewer resources during this period).

If the user does not use a disposable password or OTP solution, VPN solutions must also impose:

– rules for password refresh (every 30 days the password expires automatically);

– rules for password creation (minimum number of characters, mandatory use of alphanumeric characters).

Finally, VPN solutions manage connection criteria such as:

– the maximum number of login attempts (incorrect password) before the account is locked;

– the minimum delay between two connection attempts (the time between each attempt shall not be less than five minutes);

– the maximum number of connections that can be established simultaneously by a single user.

Of course, this application of the company's access policy should be done automatically, without any user interaction, lest the user be tempted to bypass some of the security measures to facilitate the use of IT resources. For example, they could have a connection to the Internet while connected to the business network in order to freely surf illegal download sites while waiting for results of a search on one of their company's databases. It is therefore very important for a company to carefully determine its access policy, starting with a risk analysis to allow different types of equipment (smartphone, laptop), in different configurations (with antivirus not updated) access to different resources (Web, email, application of human resources). This should be based on the principle that the most secure devices should have the greatest access, while those which have no means of protection (PDA, 3G phone, etc.) should be limited to less critical applications (Webmail).

---

11 The place of connection can be determined based on IP address, among other things. For example, the IP addresses of employees' private Internet access can be referenced, which will determine if they are working from home.

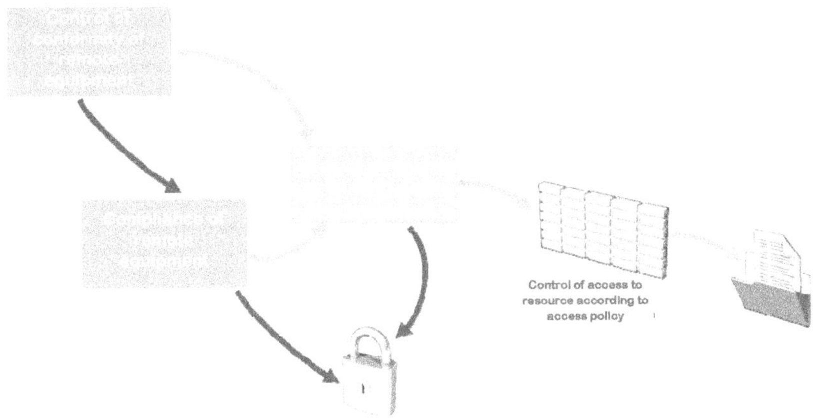

Figure 4.10. *Example of process of securing access*

4.1.4.3. *Comparison of principal tunneling solutions*

When an organization decides to implement a remote access solution, it must consider both its functional needs and the security constraints inherent to each solution technique. Therefore, in this process of selecting the best solution for your specific needs, the following summary table lists the main characteristics of the two main market solutions (Table 4.1).

	IPSEC	**SSL**
Supported client environment		
Client workstation	Business PC	Any type of client workstation (business PC, shared PC, smartphone, PDA) equipped with a Web browser. Nonetheless, if the SSL solution requires the downloading of client software, it is therefore necessary to ensure that it will function on the user's device (for example: software based on ActiveX can only be used on a Windows platform).

Table 4.1. *Comparison of the IPSec and SSL tunneling solutions*

Supported applications		
	All types of application (client server, Web) and resource (shared files).	In clientless mode, Web applications uniquely. In client modes, all types of application and resource (shared files).
Use		
Ease of use for the end user.	The IPSec client can be somewhat complex for an inexperienced user.	Use involves no more complexity than the everyday use of a Web browser.
Personalized Web portal to guide the user in use of different applications and resources.	No.	Yes.
Access from the network of a partner organization.	Requires specific access permissions in the *firewall* of the partner organization.	Because this solution is based on standard Web protocols (HTTP, SSL) no particular configuration is required in the *firewall* of the partner organization.
Access from a device that is not owned by the organization (cybercafé).	No, because it requires installation and configuration of the IPSec client.	Yes, if it uses a Web browser.
Ability to create a tunnel between two sites (for example: home of a mobile worker and an organization's headquarters).	Yes.	Yes (certain developers offer VPN SSL solution between routers.).
Implementation/exploitation		
Installation of the VPN client on the user's machine.	Requires installation and/or configuration of the IPSec client on the user's machine.	The VPN client is automatically configured and installed.
Update of the VPN client on the user's machine.	In the majority of cases, requires wired distribution.	In clientless mode: no update is necessary. In client modes: automatic, because downloaded from the VPN server.
Ease of deployment to numerous users.	Requires installation/configuration of the VPN client on all user machines.	No configuration on user machines.

Table 4.1. (continued) *Comparison of the IPSec and SSL tunneling solutions*

Implementation/exploitation		
Use of solutions from different providers.	The lack of standardization can lead to interoperability problems when using heterogeneous solutions.	Standardized solution does not present any interoperability problems if not specific client used.
Granularity of audit information.	Only connection information is available.	Detailed information available both at connection level and application level.
Security		
Authentication technique.	All types of authentication (OTP, digital certificate, etc.).	All types of authentication (OTP, digital certificate, etc.).
Encryption algorithm.	Open choice of encryption algorithm.	SSL
Granularity of authorization of application access.	Only filtering by IP address/ ports number is possible (ACL).	Possibility of constraining access to whichever applications required, using the Web portal.
Control of network admission (conformity of the machine to the security policy, etc.).	Yes, if coupled to an ESCV-type solution.	Function available from certain vendors.
Masking of internal IP addresses.	Yes, if the address translation (NAT) function is implemented, or tunnel mode is used.	Address translation provided by the SSL gateway.
Protection of IP information.	Yes, if tunnel mode is used.	No.
Masking of internal DNS names.	No.	Yes, provided by the SSL gateway.
Automatic deletion of downloaded information at the end of the session.	No.	The majority of current SSL solutions automatically delete the different information stored on the user's machine during the session, such as the Web browser history, temporary files, etc.
Compatibility with NAT function.	Yes, if tunnel mode is not used, or is used with the NAT traversal option.	Yes.

Table 4.1. (continued) *Comparison of the IPSec and SSL tunneling solutions*

4.1.4.4. *How to choose your VPN solution*

To choose a VPN solution from all the deals offered by the various vendors of security solutions, you will need to take into account the four main categories of criteria summarized in Figure 4.11.

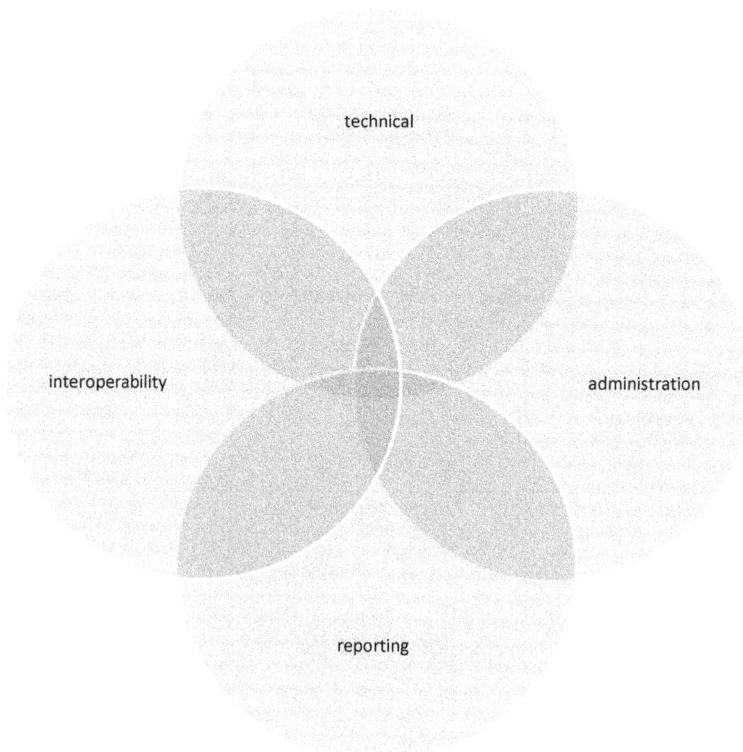

Figure 4.11. *Criteria for choice of a VPN solution*

To help you make this choice, we have detailed the essential points to take into account for each of these families' of criteria.

Technical criteria:

– VPN protocols supported (L2TP, IPSEC, etc.);

– length of encryption keys (128 bits, 256, etc.);

– authentication mode supported (TACACS, RADIUS, etc.);

– maximum number of tunnels which can be simultaneously established;

– maximum throughput attainable;

– network protocols supported (IP, IPX, NetBios, Decnet, etc.);

– routing protocols supported (RIP, OSPF, etc.);

– possibility of development (addition of cards, etc.);

– capacity for *load balancing* or high availability;

– type of WAN interface supported (Ethernet, V11, etc.);

– creation by the VPN client of a secure environment on the machine on which it is installed (automatic deletion of downloaded data, anti-*keylogger* function, etc.).

Interoperability criteria:

– operating systems supported (if using a VPN client);

– interoperability with other VPN solutions on the market;

– interoperability with different types of directory.

Reporting criteria:

– automatic creation of dashboards;

– breadth of information that can be included in the dashboard (number of connections, sharing of connections based on timeslots).

Administrative function criteria:

– simplicity of implementation and use;

– traffic management and bandwidth management functions;

– breadth of functions offered by the administration console (Web interface, configuration assistant).

NOTE.– VPN solutions are based on encryption algorithms. It is therefore important to ensure prior to implementation that they subscribe fully to the legal conditions of use in the country in which they are employed.

4.2. Control of remote access

We have seen that there are many technologies which enable an employee to remotely connect to their company's information system, through a shared network like the Internet.

Once this secure communication channel is established, before granting access to resources, the legitimacy and the "health" of the one who is on the "doorstep" of your business must be verified. This is the purpose of the following sections, which detail various aspects related to user identification/authentication, as well as the classification of information in the context of remote access.

It is necessary to implement stricter access controls for remote connections than within the physical premises of the company because:

– the devices that connect may present a higher level of risk due to their mobility. Specifically, it is possible that they have been used for non-professional purposes (surfing adults Websites) and have been subsequently contaminated, or that they do not have the latest updates (software, operating system). They may also have been stolen;

– devices connect from areas outside the perimeter of the businesses' surveillance system and are therefore not controlled nor controllable, like for example public hotspots, stations, airports, etc.

Given this context, it is vital for organizations to step up their control of remote connections to prevent an unauthorized person accessing their information systems, and also to ensure that a legitimately-established connection is not diverted for malicious purposes (propagation of worms, network scanning).

The problem of controlling access via remote connection is, in theory, equivalent to the connections made within the company LAN. In both cases, it is necessary:

– to analyze and classify the value of the different resources and data possessed by the organization;

– to define the groups of people who will have access to these classified data/resources.

However, in practice, this stage of information classification is not always carried out, or if it is, it's often carried out too vaguely, when it should be the primary issue to be resolved before being able to open your information system to remote connections with total peace of mind.

The company must analyze and classify the value of the different data types on a scale comprising at least three levels:

– non-confidential data: information that has no intrinsic value and with no consequences if disclosed (office supplies orders, service contract for the maintenance of green spaces, etc.);

– confidential information: information that can be used directly or indirectly by any person or organization and may cause limited damage (reorganization plans; orders for raw materials for production, etc.);

– highly confidential information: information which can cause significant damage if it were to be revealed (secret manufacturing techniques, marketing plan for a new product, etc.).

The servers on which these data are stored and processed must then be identified.

Finally, for each of these categories of data, the individuals or groups of employees who are authorized to have access to them must be defined, along with the modes of authentication required. For non-confidential data, a simple password authentication may be sufficient, while for more sensitive information, a fingerprint recognition system may be used, for example. It is also possible to allow access to confidential data only to those accessing it from home, and not those using a shared computer, for example.

Without this prior classification of information and identification of groups of persons authorized to access it, the technical choices of authentication mode risk being inadequate, because they will consequently not be able to achieve the desired level of security.

More generally, the lack of visibility of intangible strategic business assets and the way in which users can access them, leads to serious misunderstanding of the potential risks associated with the opening of information system to remote access.

4.2.1. *Identification and authentication*

Identification[12] and authentication[13] are directly linked processes that constitute the cornerstone of any security system. Using these mechanisms a

12 The process by which the user or the service provides a unique identifier representing his identity.

13 The process by which the user or service provides proof of his identity (shared secret, physical characteristic, etc.).

user's identity can be verified, authorizing them to access resources according to the rights they have been granted.

As we will see later, regardless of the system used to control access to resources, an authentication method is based on one of these three key principles:

– *what you know*: the user must prove to the system that he know's "secret" information (password, safe combination) by directly providing it or giving the unique result of an operation that can only be obtained by an individual who holds that information;

– *what you have*: the user must possess a particular object (key, smartcard, badge);

– *who you are*: certain characteristics unique to the user (face, digital fingerprint, voice) can be used to establish the user's identity.

Each of these methods has its own advantages and constraints. There is thus no perfect solution, so you need to choose one that best meets your needs, or use a solution based on a combination of these principles to combine the respective advantages, a method known as strong authentication.

NOTE.– It should be noted that many organizations automatically force remote users to re-authenticate if their session lasts longer than a certain number of hours, or if there is no activity for some time. This avoids the risk of an unauthorized person benefiting from a user's negligence and using their session.

4.2.1.1. *Static password*

This is one of the older authentication modes. It consists simply of asking the user, after identification, to give a password. When using this authentication mode, it is necessary to:

– define a security policy for passwords:

- duration of password validity[14],

14 This subject is debated in the community of computer security specialists. Even though most security policies include the principle of periodically changing the password, some experts are finding the effectiveness of this measure questionable, for several reasons:

- number of attempts possible before the account is blocked,

- complexity of password;

– minimum number of characters of which the password can consist;

– type of characters of which the password must be composed (alphanumeric, special characters);

– prohibition of words that can be found in a dictionary (whether or not they are in the local language);

– define a policy for reviewing account use:

- automatic locking of accounts which have not been used for a certain period of time (time to be defined),

- criteria for controlling authentication anomalies.

These rules apply in all environments, but remote access also has specific needs to be taken into account.

So, how to proceed in issuing a new password to a mobile user who has lost theirs? In this case, to prevent a hacker attempting to impersonate a legitimate user, it is necessary to implement an intermediate degraded authentication procedure. There are several possible ways of achieving this:

– ensure that the call is from a mobile phone number belonging to the user and which is declared in the company's database of GSM phones;

– ask the user to call a third party (head of department) who can formally identify him. The third party must be present on the company's premises, to prevent the hacker using an accomplice to impersonate this third party;

– develop a Web portal to reset the password after the user has responded to a number of so-called "secret questions" (mother's maiden name, the name of his first pet, model of his first car) he have previously given.

- on the one hand, techniques of "cracking" passwords have made enormous progress in recent years (see the *rainbows tables* approach). In this framework, is a policy of changing a password every three months relevant?

- on the other hand, some believe that it is more efficient to implement a very responsive monitoring device, warning as quickly as possible of abnormalities in authentication/identification and raising users' awareness with respect to reporting any problem of this kind, in return for the abandonment of periodic password changes, which by their nature are complex to remember.

The robustness of a password depends principally on three elements:

– the length of the password (number of characters that constitute it);

– the number of possible combinations for each of the characters which compose the password;

– the absence of major cryptographic weakness in the authentication system itself.[15]

As can be seen in Table 4.2, based on just the first two parameters it is possible to calculate the number of possible combinations for a password, depending on the diversity of the range of characters used.

Length of password	Characters used	Number of combinations
8	Alphabetic (a-z)	208,827,064,576
8	Alphanumeric (a-z + 0-9)	2,821,109,907,456
8	Alphanumeric with lower and upper case (a-z + A-Z + 0-9)	218,340,105,584,896
8	Alphanumeric with lower and upper case and 20 special characters (a-z + A-Z + 0-9 + &-@)	2,044,140,858,654,976

Table 4.2. *Password robustness according to characters used*

It is therefore obvious that the longer the password and greater the range of characters (alphabetic, lowercase, uppercase, numeric, special) are, the more complicated the hacker's task becomes.

15 Windows "LanMan authentication" truncates passwords up to 14 characters into two sets of seven characters, and is not case sensitive. This makes a brute force attack much easier, because the number of possible combinations is drastically reduced.

There is another static authentication solution which does not rely on the use of a password, but instead on a digital certificate.

As in real life, where we refer to documents issued by the state (identity card or passport) to prove our identity, in the digital world authentication certificates (still known as digital ID) are used. These certificates are issued by specialized "third party certification" agencies, such as Verisign, for example.

To produce this documentation, the certification authority uses the following process:

– *Creation of a certificate* which contains the information presented in Figure 4.12.

– *Calculation of a signature*, of the content of the certificate using a hash algorithm, and then encryption of the result with the secret key of the certification authority. This provides assurance that the contents of the certificate have not been tampered with and that the identity of the issuer is valid. To check this, all that is required is to use the shared key of the certification authority to verify that the encrypted summary corresponds to the certificate's signature.

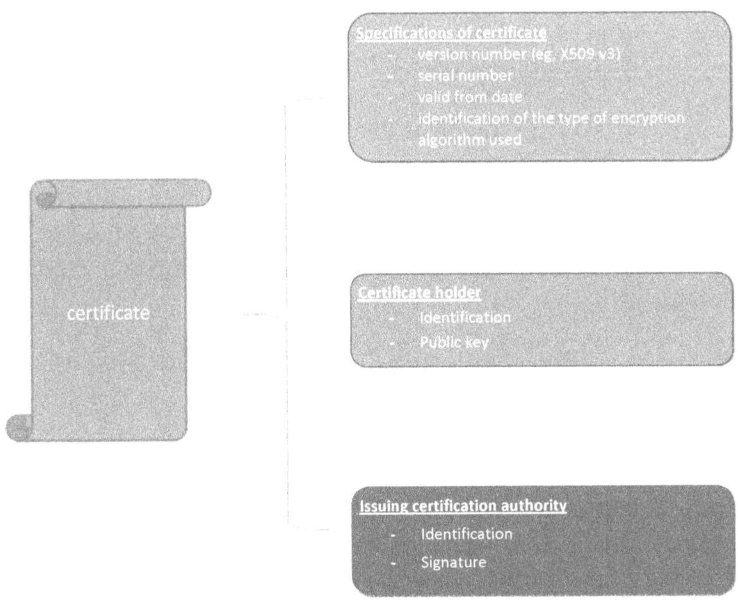

Figure 4.12. *Structure of a certificate*

For a company which uses many digital certificates, it would be too expensive to use systematically an external authority to, for example, authenticate each of its users. Therefore, companies can also implement their own *Public Key Infrastructure* (PKI), which consists of three main modules:

– *Registration Authority*, which is responsible for managing certificate requests made by users, and their submission to the certification authority, once the identity of the applicant has been verified;

– *Certification Authority*, which generates the certificate and signs it with the secret key;

– the central repository, which stores certificates so that they can be consulted, and which also manages revocations.

As part of deploying a remote access solution, it is strongly recommended that mutual client/server authentication is used when establishing a connection. To do this, the remote access server provides the client with its digital certificate to prove its identity before asking the user to authenticate; in this way, both parties are certain of the other's identity.

4.2.1.2. *Dynamic password*

A dynamic password system[16] is based on the principle that the authentication code provided is only valid once (that is, disposable). With each new connection request, the user must generate a new code to be provided to authentication system.

This type of system combines the principles of "what you have" and "what you know". The user must therefore possess an external device which constitutes the "what you have". This is usually in the form of a small box (calculator, keyring) that automatically generates a series of numbers, which are only valid once, but which could equally be a cellphone which receives an SMS code.

There are two different techniques for generating this series of numbers:

– *synchronous mode*: the authentication server and the token automatically generate a code at regular intervals (usually 30 seconds), based on time (date, time of day), certain events (number of connection attempts) and information related to the user (shared secret). The user adds to the set of numbers generated a PIN (*Personal Identification Number*) code which is personal to him, and he has chosen himself. Concatenating the user's PIN

16 Also known as *one-time password*.

and the figures provided by the equipment gives the final authentication code. This system has notably been used by RSA *Security* for their *Secure* ID product;

– *asynchronous mode*, based on the challenge-response concept: the server generates a code (stimulation), which acts as a PIN, which the user enters into their token. It then calculates a response from the code and confidential information it shares with the server (shared secret). The user simply uses the result to authenticate.

The PIN code used in synchronous and asynchronous modes is the "what you know" which, coupled with the "what you have", constitutes a strong authentication system.

This method most notably offers protection against key logger attacks and sniffing because the passwords are disposable. However, although the level of safety obtained thanks to these devices is high, attacks are still possible[17,18].

It should be noted that with the development of GSM phones, smartphones such as BlackBerrys, an increasing number of companies have abandoned the use of traditional tokens in favor of sending passcodes directly to mobile devices, for example using SMS messages. This allows them to reduce their costs because they no longer have to acquire or manage specific proprietary equipment to ensure the delivery of a One Time Password.

The use of dynamic authentication based on specific equipment is not unique to remote access and can also be used internally in an organization. Nevertheless, in practice this device is deployed primarily with mobile users to ensure that the person being authenticated is a certified member of staff.

4.2.1.3. *Authentication servers*

4.2.1.3.1. AAA Servers (Authentication, Authorization, Accounting)

Once the size of an organization becomes significant (a hundred accounts and applications), it is necessary to use authentication servers to centralize

17 A hacker can attempt to steal a user's token and obtain their personal code using the social engineering technique.

18 In March 2011, RSA, producers of SecureID equipment, were compromised following a targeted attack. This allowed hackers to break into the information system of Lockheed Martin, whose employees used SecureID tokens.

the management of authorizations. For this, authentication servers use databases to store users' access rights. These data can either be integrated directly into the authentication server, or – more commonly – stored in one or more directories[19] that the authentication server consults. The acronym AAA (*Authentication, Authorization, Accounting*)[20] is generally used to refer to this type of architecture, representing the three basic functions of an authentication server. It should be noted that historically the AAA server deployment emerged from the need for Internet service providers to manage the users connecting via a modem to their infrastructure, in order to bill them for connection time to their services. Although currently, features and technical architectures have evolved into access packages based on usage, the need for an accounting function remains, but it is primarily used to track and detect suspicious behavior during remote connections.

If the principle of the AAA architecture is unique, solutions for deploying it are numerous. We chose within the scope of this book to limit ourselves to the three most prevalent technologies: RADIUS, TACACS and Kerberos.

4.2.1.3.2. RADIUS (Remote Authentication Dial-In User Service)

RADIUS is a system for authentication as well as management of authorizations and accounting, which is based on a client-server architecture type and UDP protocol (port 1812). It was developed by Livingston (which was later acquired by Lucent) and normalized in 1997 by the IETF (RFC 2865[21]).

When a user wishes to connect, he sends his request to a RADIUS client (*firewall*, VPN appliance, Wi-Fi access point), which is responsible for collecting the information necessary for identification/authentication. Once

---

19 Directories are databases containing the details of each user of the information system with a certain amount of associated administrative information. This is the single repository that the various components of the IS will query to obtain the data concerning these users. Novell, with NDS (*Novell Directory Services*), popularized the use of directories to administer PC networks. The idea was taken up by Microsoft with *Active Directory* which was integrated in the basis of Windows 2000. However, the real explosion in the use of this type of solution came with LDAP (Lightweight Directory Access Protocol).

20 In the literature, you may also find the concept of WWW, which is equivalent to AAA, in that the acronym signifies:

– *Who are you?*
– *What are you allowed to do?*
– *What did you do and how long did you do it for?*

21 Replacing RFC 2138 and 2139.

this task is achieved, it issues an authentication request (*Access Request*) to the server. The RADIUS server then consults an identification/authentication database to compare the information contained therein with those of the query. Depending on the outcome, the server will either:

– ask the user for additional information (*Access Challenge*);

– refuse the connection (*Access Reject*);

– accept the connection (*Access Accept*).

The server returns the connection profile to the RADIUS client in the Access Accept packet, allowing the RADIUS client to know under which conditions the user is permitted to connect.

One of the critical points in terms of safety with RADIUS is transmission of identification and authentication data between the client and the server, because the entire content of the Access Request packet is transmitted unencrypted (with the exception of the password field which is the subject of an MD5 hash calculation based on a shared secret between the client and the RADIUS server). Due to the use of this method, several types of attack are possible, if the communications between the RADIUS server and the client are accessible to an attacker. It is therefore recommended to protect communications between the server and the client using encryption (IPSEC), or to use a secure link guaranteeing the prevention of eavesdropping by an attacker.

It is also worth mentioning that using a strong authentication system (see section 4.2.1.2) with a RADIUS server limits the attacks described above, since the knowledge of the shared secret between the server and the client does not allow replay of dynamic passwords.

4.2.1.3.3. *TACACS+ (*Terminal Access Controller Access Control System*)*

TACACS+ allows authentication as well as permissions management and *accounting*. It is based on a client-server architecture type, and uses TCP, unlike RADIUS. Originally, TACACS was developed by BBN Planet Corp., for the U.S. Department of Defense. Today, Cisco continues to develop it.

Its main advantage over RADIUS is that it encrypts all credentials (username, authorized services, etc.) which traverse the network and not just the password, thus preventing interception of sensitive data.

4.2.1.3.4. Kerberos

Kerberos (name of the three-headed canine guardian of the gates of Hades in Greek mythology) is an authentication protocol developed in 1983 by MIT as part of the Athena Project, which aimed to design distributed computing environments. It has also been standardized by IETF in RFC 1510.

Compared to the two solutions presented above, Kerberos has the distinction of being based on a system of exchanging tickets rather than passwords for authentication and access to various IT resources.

When a user wishes to access an application, they first send an authentication request to the AS (*Authentication Server*) as shown in steps 1 and 2 of Figure 4.13. The SA then returns a TGT (*Ticket Granting Ticket*) that cannot be decrypted except with a password, thus avoiding that the password is transmitted on the network. Once this operation is achieved, the user will use the session key contained in the TGT (replacing the user's password) to request a service ticket from the TGS (*Ticket Granting System*) server: steps 3 and 4 in Figure 4.13. When the service ticket is issued, the user can then use it to connect to the application: steps 5 and 6 in Figure 4.13.

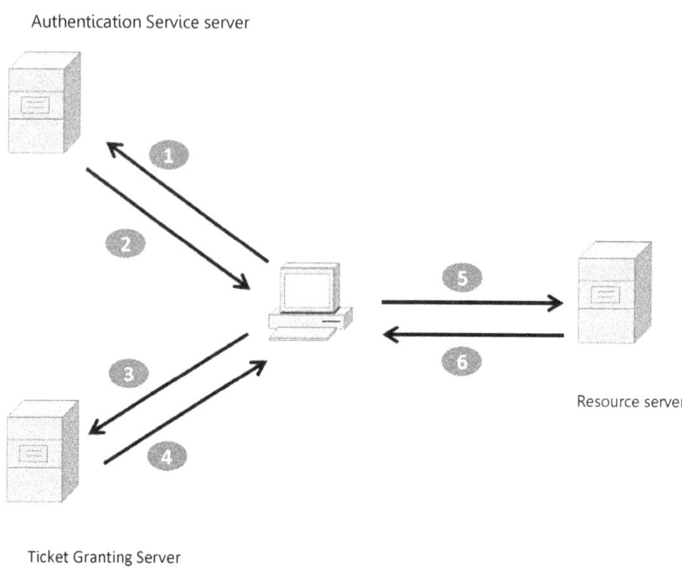

Figure 4.13. *Kerberos architecture*

This solution was chosen by Microsoft as the default protocol for authentication performed in the Windows domains (*Active Directory*).

Kerberos is one of the solutions used to implement an SSO and prevent the transmission of *password/login* over the network, since it relies on a limited duration ticket system. It has many advantages, including being adopted by many operating systems (Linux, Windows).

However, Kerberos has limitations and disadvantages, the main ones being:

– it requires a significant investment to master its operation;

– the compromise of a single authentication server compromises the entire architecture.

4.2.1.4. *Securing of smartphones/tablets and BYOD (Bring Your Own Device)*

Historically, the term *smartphone* designated a mobile phone which also featured calendaring and contact management functions. But in 2007, with the launch of the iPhone, Apple revolutionized the market by offering a device that was a fusion of a mobile phone and a computer.

Today, Apple is still a major player in this market (23.8%), although it faces strong competition from Android systems (50.9%).

Smartphones pose a challenge for security teams due to their size and their functions (phone, calendar, contact management, etc.). They travel with their users at all times and can easily be stolen. And as we have seen, there are now real computers[22] that are subject to the same risks.

In addition, a new phenomenon called Bring Your Own Device (BYOD) emerged in the United States. This concept refers to instances where employees use their own equipment to work on the tasks entrusted to them by their company. This does not apply just to smartphones/tablets, but also includes laptops and in general any personal equipment that an employee uses in connection with their job. The threats posed by BYOD were discussed in section 2.6.

---

22 The majority of smartphones use a Unix system (Android and Apple's iPhone, with RIM also soon to move towards such an operating system with QNX).

Worldwide Smartphone Sales to EndUsers by Operating System in 4Q11 (Thousands of Units)				
Operating System	**4Q11 Units**	**4Q11 Market Share (%)**	**4Q10 Units**	**4Q10 Market Share (%)**
Android	75,906.1	50.9	30,801.2	30.5
iOS	35,456.0	23.8	16,011.1	15.8
Symbian	17,458.4	11.7	32,642.1	32.3
Reasearch in Motion	13,184.5	8.8	14,762.0	14.6
Bada	3,111.3	2.1	2,026.8	2.0
Microsoft	2,759.0	1.9	3,419.3	3.4
Others	1,166.5	0.8	1,487.9	1.5
Total	**149,041.8**	**100.0**	**101,150.3**	**100.0**

Table 4.3. *Evolution of the smartphone market ©Gartner*

We are now experiencing a revolution: the breaking down of barriers between professional and private life. Clearly, the BYOD phenomenon has many security implications.

Previously, a user could use their work computer for personal use, but only if it met the acceptable use policy imposed by the company (e.g. prohibition of playing online games). It was therefore this last issue which set the "rules of the game". With BYOD, the employee's personal device serves a professional purpose, so the rules are reviewed to suit. For example, while a company could legitimately destroy business data on a device they owned, they must now obtain permission to do so on equipment owned by their employee.

Smartphones now have vast libraries of applications that are difficult for the IT department to control. A simple search on Apple's *Store* brings up a network scanner, anonymization systems (*tor browser*), Skype, IRC clients, etc.; in other words, enough applications to allow all kinds of information leakage. Furthermore, these information leaks are not necessarily voluntary, since many users are tempted to download apps from alternative *appstores* which, unfortunately, are full of trojans.

The catalog of security products (antivirus, *firewall*, etc.) is much less mature for smartphone environments than for laptop computers.

Smartphones are extremely diverse in terms of hardware type (iPhone, BlackBerry, Galaxy, etc.) and it is not possible to guarantee that these devices have the latest versions of the operating system or the latest patches, since their installation depends on users' enthusiasm for doing so. This is a veritable nightmare for the services in charge of security, because an un-updated system may have numerous loopholes which can be exploited by hackers.

To address these threats, your organization has the choice between:

– simply banning the use of personal devices. In this case, you must deny these devices access to corporate resources. To do this, there are many technical solutions, such as blocking access to Web Intranet servers based on the user agents[23] of browsers used by smartphones, or using digital signatures[24] to authenticate approved devices;

– accepting the use of personal devices for professionnal purposes while at the same time reducing the security impact on the information system. In this case, you should follow the methodology shown in Figure 4.14.

23 A *user agent* is a chain of characters which identify the navigator used, its version, and it operating system.

24 There are solutions which allow the generation of a "digital ADN" for each device; in other words, an identity card based on material which will identify it in a unique manner. The organization can therefore ensure that the devices that connect to their systems are those provided by the firm.

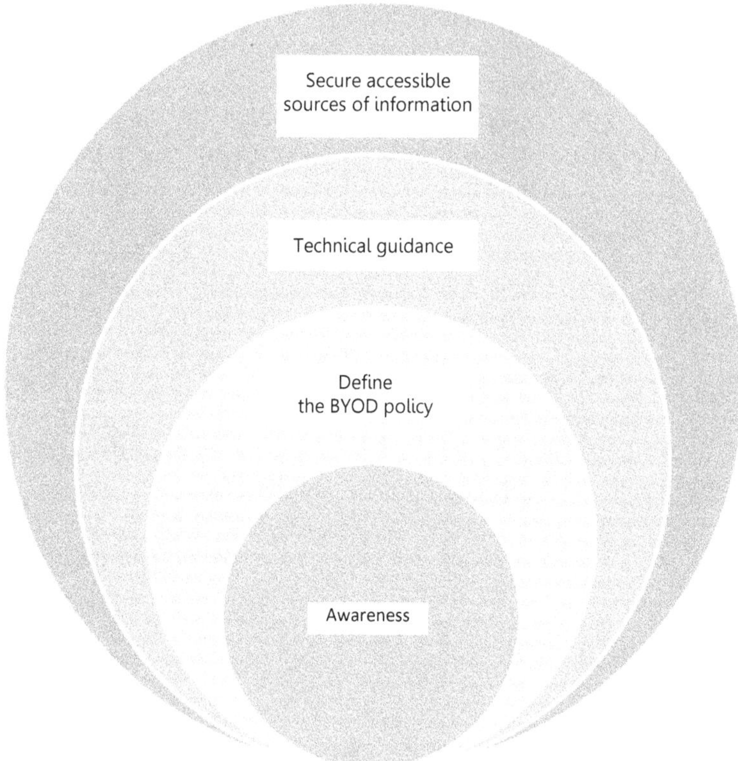

Figure 4.14. *Steps necessary for implementing a BYOD policy*

a) Awareness

- the first line of defense for your IT system is the users themselves. It is therefore essential that they have the knowledge necessary to understand security issues in private and/or professional contexts. This is particularly important in the case of BYOD, because there is a huge disparity in the level of use of smartphones' various functions by different users;

- it is crucial to pay attention to the quality of education, which should be adapted to the target audience. Too often this type of training activity is over-long and does not capture the attention of the audience, limiting its effectiveness;

- this training must go hand in hand with a formal definition of roles and responsibilities in the use of BYOD, as we will see.

b) Define the BYOD policy

- it is necessary to establish a clear BYOD usage policy which specifies what is allowed and what is prohibited in the context of the use of personal equipment. This is a crucial point which will prevent misunderstandings between users and technical/security;

- this BYOD policy should be incorporated into the computing resource usage policy, which users must sign;

- to the extent that these devices mix private and professional life, it is necessary to involve the legal department and human resources in the drafting of this document, because there are many sensitive issues to be addressed. For example, the user may authorize tracking of the phone's signal in order that the device's location can be traced, and the device recovered in case of theft. Or, they may allow the organization to remotely delete (*wipe*) business data stored on their personal smartphone.

c) Technical guidance of users.

It will be necessary to technically supervise the use of personal devices in the professional context, to:

- on the one hand, prevent your employees finding their own solutions that work, but that do not have an adequate level of security. For example, if you do not tell them how to configure the smartphone's *native email* client to retrieve emails, they may be tempted to connect *via* an integrated browser, the configuration of which does not correctly remove cookies after logout. They could also choose their own chat client to access instant messaging, which would not provide the same security guarantees as a tool approved by your technical staff;

- on the other, be able to install security solutions chosen by the company (antivirus, firewall, encryption solution, remote system cleaning – wiping the smartphone in case of theft, etc.).

d) *Secured sources of* information accessible by your mobile users, particularly via their smartphones. Despite all precautions, the user's device may still become compromised. Therefore, as with any project involving opening up an information system to remote access, it must be ensured that sensitive company data have been identified and that there is an adequate level of protection which takes into account the new threats that could be introduced by this new use. For example, data stored in the mobile device could by routinely encrypted so that in case of theft, the data cannot be exploited.

As you can see, these measures are taken mainly at the organizational level to supervise, educate and assist your staff in the use they will make of their mobile device. Technical security solutions are certainly essential, but their efficacy can only reflect the prior organizational efforts made.

To help you in the choice of technical solutions, we have summarized in Table 4.4 the various existing solutions for securing a smartphone/tablet.

	iOS	**Android**
Firewall	No[25]	Yes
Antivirus	No[26]	Yes
Data encryption	Yes	Depends on version + third-party applications
Wiping system	Yes, by design	Yes, with third-party applications
Enforced use of complex passwords	Yes	Yes
Mandatory locking of smartphone	Yes	Yes
Locking of update sources	By design	No
VPN system	Yes	Yes
Prohibition to change profiles Configuration	Yes	Function not available
Deactivation of certain applications[27] when the user connects to the organization's network	Yes	Function not available

Table 4.4. *Security solutions available on the main types of smartphone*

25 A firewall application exists but requires the iPhone to be jailbroken, which is not compatible with organizations' security policies.

26 Antivirus software exists today but with limited capabilities compared to a classic PC (no scan of installed applications or file system). To date, no major vendors of antivirus software offer products for iPhone.

27 Solutions exist for disabling certain native smartphone applications including social networks, or creating specific profiles via the application manager which can force its installation (push) on the smartphone when it connects the corporate network.

As can be seen, the technical solutions deployed on smartphones and tablets strongly resemble those that can be found on laptops (encryption, firewall and antivirus).

The main source of insecurity and difficulty remains the mix of personal and professional data, with the related risks of data leakage (accidental or not) linked to the extreme mobility of these types of device.

In fact the first and most important defense stems from staff awareness, a line of defense complemented by the protection of sensitive sources of information (strong authentication, encryption, access control, etc.).

It is important to note that this type of equipment does not require a revolution in the means developed for protecting and securing them, but simply the definition of a strict framework for their use.

4.2.2. *Unique authentication*

In large organizations, the number of applications used by employees has increased dramatically in recent years. As a result, the number of passwords that users must manage has followed the same trend.

Too much security harms security, and one of the major consequences of this explosion in the quantity of passwords has been the emergence of risky behavior (recording the password in a notebook or using passwords which are too simple, etc.). In addition, security policies requiring regular password changes (with storage of previously-used passwords for periods of six months to a year), are sometimes applied somewhat indiscriminately (with no support to help users manage their numerous passwords) and have amplified these risky behaviors (see Figure 4.15).

For this reason, SSO (Single Sign On) solutions emerged a decade ago, providing an interface between the different authentication systems and the end user so that they can access resources only having to enter their password once.

The advantages of implementing an SSO are diverse (see Figure 4.16).

Figure 4.15. *Ecosystem of Web users (source: BT IAM department)*

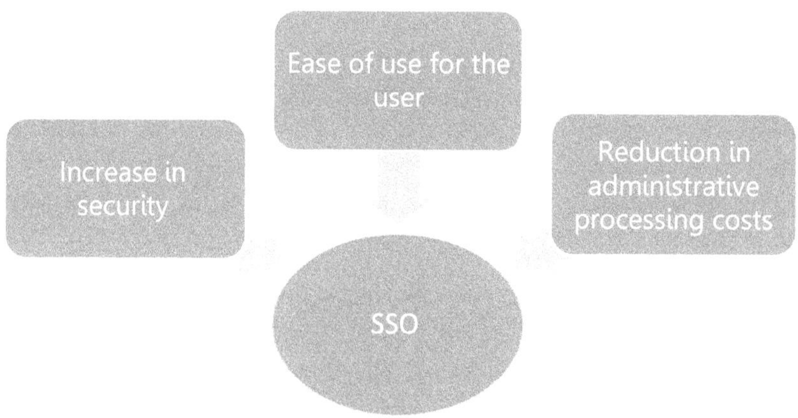

Figure 4.16. *The benefits of SSO*

Firstly, with regard to security, the establishment of a SSO principally offers:

– improved traceability of users via unique authentication;

– reduction of risky behavior of users associated with multiple logins/passwords.

It will also reduce the cost of administrative processing, due to a decrease in the number of passwords resets associated with them having been forgotten, and interventions to unlock blocked accounts after multiple connection attempts. These operations are estimated to have an average cost of $30 per person according to the Meta Group.

Nevertheless, this type of solution can be difficult to implement and requires modifications to certain applications, or to develop specific patches to ensure that existing authentication systems will interface with the SSO software.

For this reason, we are currently witnessing the development of two protocols based on XML to enable different authentication systems to exchange data with the aim of federating identity management. The first SAML (*Security Assertion Markup Language*) is supported by the association Liberty Alliance, while the second, *WS-Federation*, is supported by IBM and Microsoft.

4.3. Architecture of remote access solutions

As mentioned in the introduction to this chapter, the design of specific architectures for remote access is a major factor in all technical security measures. An error in the design of this infrastructure could expose your entire information system. Consequently, this section discusses the different architectures for implementing a remote access gateway.

4.3.1. *Securing the infrastructure*

To reduce risk, it is strongly recommended that you do not connect the remote access server to the Internet, but instead install it behind a firewall in a DMZ. Thus, in the case of an attempted attack, the attacker must first thwart the security mechanisms of the *firewall* before attempting to corrupt the remote access server. Even if the hacker managed to compromise the latter, it is located in a DMZ, and they would therefore find it very difficult to use it to access the organization's information system, because the *firewall* only accepts the establishment of connections between the VPN gateway and servers/resources authorized for remote access use.

It is even more important that the data traveling towards the Intranet are passing through a firewall because as data are encrypted until they reach the VPN gateway, the security devices are not able to analyze them and therefore to detect any threat they contain. It is only once the data "exit" from the VPN gateway that the *firewall* can fully use its analysis capabilities.

As you may have guessed, we can further reinforce this architecture by using not one, but two separate *firewalls*: one to protect access to the Internet and the other to defend the company's network. Preferably, choose two different vendors for both *firewalls*. Thus, if an attacker can exploit a security flaw in one of the two *firewalls*, it will be difficult to repeat this feat with the second.

This architecture may also be supplemented by implementing one or more relay servers (proxies) in the DMZ, so that remote users never have direct access to the company's resources. This will also allow the company's internal DNS structure to be "hidden", to further complicate the task of potential hackers (see Figure 4.17).

And since you can never be too careful, an IPS can also be implemented to analyze the various flows that pass through the DMZ, looking for threats such as viruses hidden in a file. As we saw in the previous chapter, *firewalls* and proxies do not use the same methods of traffic analysis as IPS and it is highly recommended that they should be used complementarily. Of course, as already explained, the IPS should be placed downstream of the VPN gateway to be able to analyze such traffic.

Finally, it is important to carefully monitor the resources that the remote user can access and not to open up the entire internal network. As we have seen, this is easily achievable with a remote access SSL server. But even with the IPSec solution, access control lists (ACLs) can be used to limit the potential scope of an attack and only allow connection to resources which are strictly necessary.

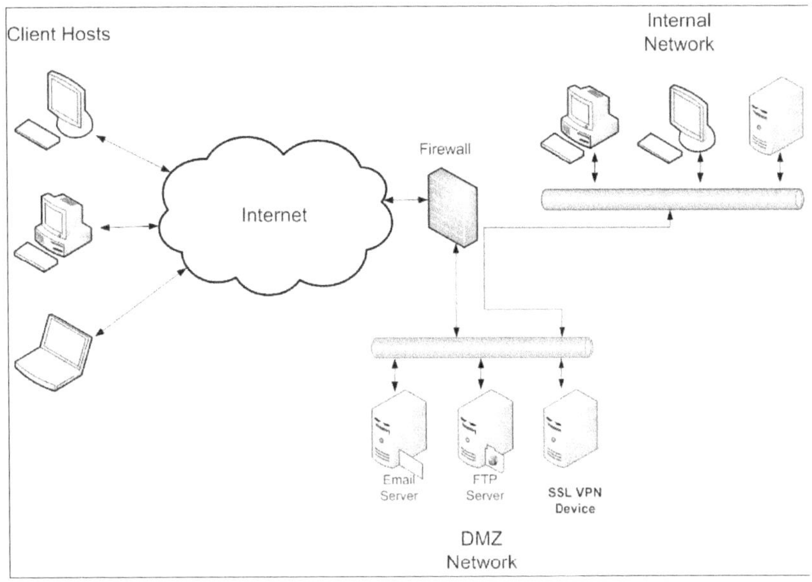

Figure 4.17. *Example of the insertion of the VPN gateway into the DMZ* ©*NIST*

NOTE.– To avoid creating as many connection profiles as users, which inevitably result in great complexity in daily management, it is preferable to define several families (accounting, R&D, security, etc.) with access corresponding to their functional needs and to associate client profiles with these.

Of course, all these access events and attempts will be recorded in a separate log server to help consolidate the different events and detect an attempted attack, and also to allow a *post-mortem* analysis to be carried out after a security incident to discover how the hacker managed to circumvent protection measures, in order to establish an action plan to address these deficiencies.

This section would not be complete without the mention of alternatives to the installation of the VPN gateway in the DMZ, and their respective advantages and disadvantages (Table 4.5).

It should be noted that an increasing number of vendors offer integrated solutions including VPN concentrators, *firewall*, antivirus, *anti-spyware* and IPS, avoiding deploying and managing multiple security devices. The main advantage of this architecture is that as security solutions are integrated into the same housing as the VPN gateway, they can directly analyze flows as soon as they are decrypted.

However, the main drawback of this solution is that as encryption/ decryption of the VPN gateway are very consuming in terms of computer resources, they can impact the performance of security applications and therefore provide the end user with a degraded service in terms of response time.

Architecture	Drawbacks	Advantages
VPN gateway inside the DMZ.	- Requires opening numerous access points in the *firewall* between the gateway and the intranet. - Flows to servers located in the DMZ are not controlled by the firewall, because they are encrypted.	- Streams are decrypted before being sent to the intranet and can be controlled by the *firewall.* - The VPN gateway is protected by the *firewall.* - If the VPN gateway is compromised, the hacker will not have access to the entire intranet as he is blocked by the *firewall.*
Gateway connected directly to the Internet, in front of the *firewall.*	- The gateway is not protected from attack coming from the Internet. - Requires opening. numerous access points in the *firewall* between the gateway and the intranet.	- Streams are decrypted before being sent to the intranet and can be controlled by the *firewall.* - If the VPN gateway is compromised, the hacker will not have access to the entire intranet as he is blocked by the *firewall.*
Gateway connected directly to the intranet, behind the *firewall.*	- The *firewall* cannot control flows, because they are encrypted. - If the VPN gateway is compromised, the hacker has access to the entire intranet.	- The VPN gateway is protected by the *firewall.* - Only requires the opening of one access point in the *firewall* between the exterior and the VPN gateway.

Table 4.5. *Advantages/disadvantages of different VPN gateway architectures*

4.3.2. *Load balancing/redundancy*

In the literature, the different functions which must be covered by security measures are often classified into three main areas: integrity, confidentiality and availability.

As you have understood from this chapter, the confidentiality and integrity of data transmitted over the network, are ensured by the intrinsic functions of a VPN.

But it is not the same with availability. Thus, for the VPN solution to continue to provide its service in the event of technical failure or a hacker attack, it is essential to implement a redundant architecture with or without load balancing.

Of course, this reinforcing of infrastructure does not just concern security devices (VPN gateway, authentication server, etc.), but also all other components of the communication channel (routers, switches, DNS server, etc.) because, as the saying goes: "A chain is only as strong as its weakest link".

There are two main types of implementation of the high-availability infrastructure, based on the use of at least two devices of the same type, an active/passive mode and an active/active mode.

To explain the principle, take the case of an architecture consisting of two servers:

– active/passive mode: the first device is known as the master and ensures all requests are processed. The second server is called the slave and remains passive (on standby) until the master is no longer able to perform its function (failure, maintenance, etc.), at which point the slave takes over;

– active/active mode: both servers provide processing of all applications "fairly", according to a predefined algorithm (Round Robin, etc.). When one of them happens to be unavailable, the other takes over the processing of all requests.

Based on implementations from different manufacturers, changeover in the case of unavailability of one of the devices may or may not be transparent to the user. Because for a changeover to be achieved, it is necessary that for servers to constantly exchange information regarding the processes they are performing, which can be very complex to achieve.

Some solutions even allow up to 255 machines in a cluster.

A cluster allows:

– better availability of service: in case of the failure of one of the devices, the customer is taken care of automatically (with or without interruption of the session) by another machine. Similarly, in the case of maintenance of one of the devices, the service will continue to be provided, because the other machines will not be stopped;

– ensure the scalability of the solution: because processing demands are distributed across all the machines, many clients can be handled. In the case of saturation of the capacity of the machines, a new device can simply be added to the cluster to increase the global capacity of the solution;

– simplify the management of different machines: most cluster management solutions seamlessly manage all the machines composing the cluster as if there were only one. For example, updating a configuration will automatically be applied to all equipment without requiring special intervention on the part of the operator.

There is an alternative to the clustering solution for large international companies, which consists of installing a VPN concentrator in each major region around the world where the company operates. For example, in case of failure of European access, users can use the Asian facilities. The only impact of this is a degradation of response time, due to the longer distances travelled by IP packets between the client device and the VPN gateway.

4.4. Control of conformity of the VPN infrastructure

As we have seen, the VPN gateway is a cornerstone of the security of the information system. If the remote access infrastructure is compromised, the entire resources of the company are exposed to potential attacks from the outside.

It is therefore essential to use best practice with these devices in terms of management of security risks:

– installation of software patches in the shortest possible time;

– limiting access to authorized individuals only, and if the solution allows implementation of user profiles with limited rights, only to functions essential to performing their duties;

– limiting the number of computers allowed to connect to these devices to perform administrative tasks, etc.

To ensure that the elements constituting the VPN infrastructure meet these guidelines and thus contain no security vulnerabilities that could be exploited by a potential attacker, conformity audits must be carried out.

Of course, such audits should be carried out regularly, as new vulnerabilities are discovered daily, and production environments are constantly evolving (adding new rules, installation of a new version of software, etc.) thereby creating new vulnerabilities.

To evaluate the conformity of the infrastructure in its entirety, you can use external consultants, via PenTest audits, or if you have people in your team with a sufficient technical level, use software for scanning network vulnerabilities.

These programs check that only the services you wish to offer are available and no others, and moreover that they do not contain flaws.

One of the first tools developed for this purpose was the SATAN (*Security Analysis Tool for Auditing Networks*) software developed by Dan Farmer and distributed as open source on the Internet since 1995.

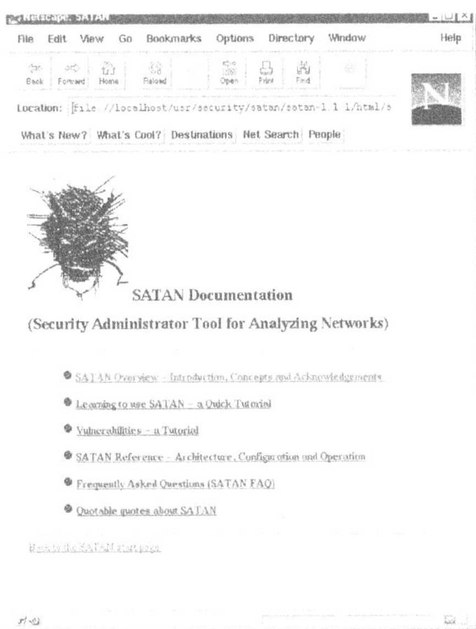

Figure 4.18. *SATAN software*

Since then, many other programs have emerged, either from the public domain (Nmap, Nesss, etc.) or marketed by vendors (NetRecon, NetSonar, Qualys, etc.). In either case, even if the interfaces differ and present greater or lesser ease of use, all these tools generally have two modes of operation:

– a mode for collection of information on the different components of the information system (routers/switches, firewalls, operating systems, applications, Web servers, etc.) from the network to detect potential risks associated with the use of an older software version or configuration faults, and then suggesting a plan of action to eliminate them (installation of patches[28], modification of security settings, etc.).

– a mode for simulating the behavior of a hacker and actual implementation of attack attempts. Of course, the latter mode must used with great care as it can cause unintentional malfunction.

If you wish to acquire this type of tool, a number of criteria should be taken into account in order to choose the product that best suits your needs.

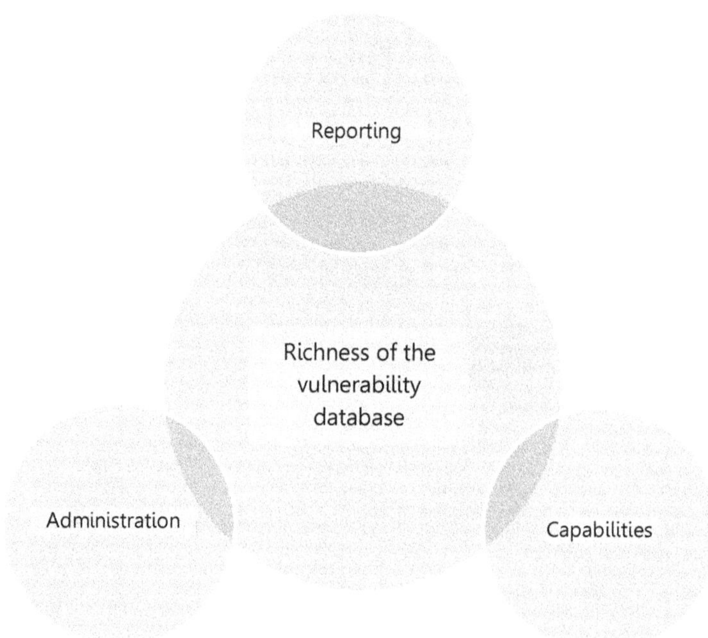

Figure 4.19. *Criteria for choice of network vulnerability scanner*

28 Corrective resolving programming errors in a piece of software.

Richness of the vulnerability database:

– threats evolve very quickly, it is essential that the vulnerability database used by the tool to perform its scans is updated regularly by the vendor to take into account the latest vulnerabilities;

– faults that the tool can detect must be covered by the main nomenclatures which reference vulnerabilities: CVE (Common Vulnerabilities and Exposures), CERT and Bugtraq ID;

– the tool must be able to describe exactly what is being tested so that, should the need arise, it is possible to ensure that the reported alerts are not false positives.

Reporting:

– audit reports and recommendations should provide detailed explanations and recommendations for "filling" the gaps detected;

– it must be possible to tailor reports according to intended audience (security expert, Chief Operating Officer, etc.);

– it must be possible to produce reports in multiple formats (Word, PDF, HTML, etc.).

Administration:

– the tool must allow the search to be filtered for gaps, and to remove certain gaps when they correspond to no equipment/software installed on your information system. This will save time and avoid potential false positives;

– updating the vulnerability database (downloading by secure channel), whether manual or automatic, must be easy to achieve;

– it must be possible to schedule scans on a regular basis (weekly, the first Tuesday of each month, etc.);

– the tool must be able to generate warning messages (email, SMS, SNMP trap, etc.) when a new vulnerability is detected.

Capacities:

– time required for the tool to perform an audit of your network (for example, the Retina software from *eEye Digital Security* is able to scan a class C network in around 15 minutes);

– ability of the tool to interact with other pieces of equipment to detail corrections to be applied (system patches, modification of rights, etc.).

4.5. Control of network admission

Solutions for network admission control fall into two broad categories:

– those which restrict access to the organization's LAN to devices that have been authorized (802.1x);

– those which restrict access to the organization's LAN to devices that have been authorized (802.1x) and also ensure that their configuration conforms to the security policy in force (ESCV);

As we saw in Chapter 1, an employee may involuntarily be the vector of an attack when he connects from his workplace to his company's network if his laptop had been infected during a stay outside the company (using the Internet in his hotel room during a business trip, etc.). For this reason, in addition to measures to control remote access (see the previous section), we can implement mechanisms which control LAN access.

A study by Datamonitor, conducted in the first quarter of 2006 in North America and Europe, showed that almost 20% of respondents had experienced partial or total unavailability of their information systems due to the connection to their network of a piece of equipment which did not comply with the company's security policy.

4.5.1. *Control of network access*

As we have seen, mobile and remote access devices are at increased risk (contamination by malware, connection attempt by a hacker) compared to stationary computers that remain on the premises of the company. It is therefore recommended to implement network access controls to prevent an attempted attack bring propagated in the information system of the company.

This control can be achieved by using the 802.1x protocol or the implementation of ESCV (*Endpoint Security Compliancy Verification*) solutions presented in more detail in this section.

The IEEE 802.1x protocol controls who accesses the LAN of the company and how (management of service quality, QoS, VLAN, ACL, etc.). Figure 4.20 schematically describes the function of this protocol.

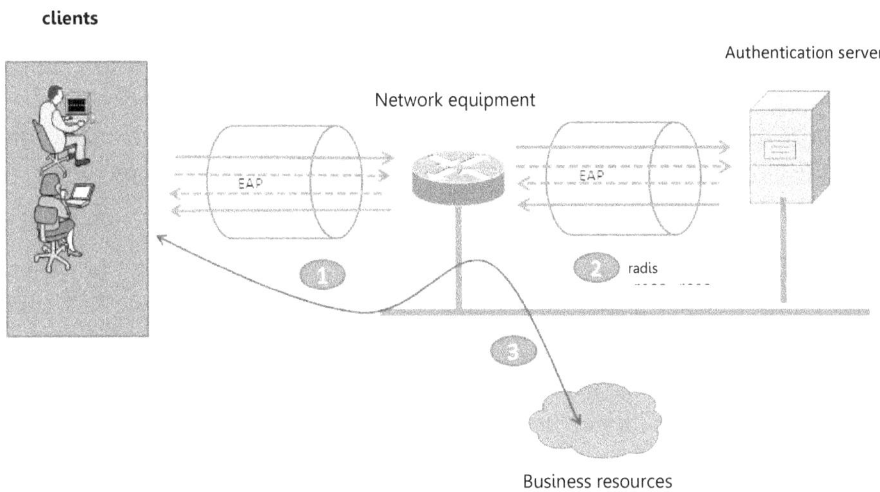

Figure 4.20. *Operating principles of the 802.1x protocol*

Thanks to the EAP protocol, the client can be authenticated through an AAA server (RADIUS etc.) (steps 1 and 2 in Figure 4.20). Once this authentication phase is successful, the switch receives from the server the access policy which principally includes information relevant to the assignment of the client to the correct port/VLAN allowing him to effectively connect to the network (step 3 in Figure 4.20). For as long as the authentication is not successful, only EAP packets are allowed to pass.

4.5.2. *ESCV (Endpoint Security Compliancy Verification)*

Control by the 802.1x protocol allows authorization of machines wishing to connect to the LAN. However, this protection does not check if the machine meets the standards set by the security policy of the company (updated antivirus software, updated security patches, etc.). However, mobile users have created new security issues, to which the simple 802.1x authentication cannot provide an answer:

– how can we ensure that machines which have been outside the company are not infected with a virus?

– how can we ensure that the connecting machine is indeed the original device provided by the company to the employee?

This prompted the development of the ESCV (*Endpoint Security Compliancy Verification*) solution which, when a computer tries to connect to the network, verifies compliance with your security policy (level of patches, active firewall, antivirus etc.). In the case of confirmed non-compliance, the machine is immediately and automatically quarantined to control and update it.

An ESCV system is based on several "building blocks" providing specialized functions that can come from different suppliers. A typical architecture consists of:

– a module that analyzes compliance with the security policy of the company of the devices connecting to the network; for this it can either be based on specific software that is installed on the equipment and which will be in charge of conducting the audit phase (version of installed patches, presence of antivirus software, latest update of anti-spyware database); or it can analyze from the network (open TCP ports). Of course, a solution based on the use of software installed on the device will provide a much more comprehensive analysis of the state of the machine;

– a communication protocol allowing all of the constituent elements of the "ESCV" to communicate with each other (state of the processed equipment, actions to be triggered following the detection of non-compliance);

– a server that centralizes the definitions of different criteria to judge the conformity of equipment, and based on the results obtained during the audit phase of the equipment, initialize the necessary actions (quarantine, disconnection of workstation, partial or total access authorization);

– a module for processing the nonconformities of the devices discovered during the analysis phase.

It will thus support:

– decontamination of infected files (viruses, worms, etc.);

– reactivation of programs (antivirus, firewall, etc.) that may have been stopped by the malware to allow it to carry out its function;

– uninstalling unauthorized software (peer-to-peer, etc.);

– updating installed software (signatures database for antivirus software, etc.) and the operating system (patch);

– installation of new security settings (new rules for the firewall).

In some cases/solutions, such compliance is not automatic: the user or a member of the technical team will make these corrections.

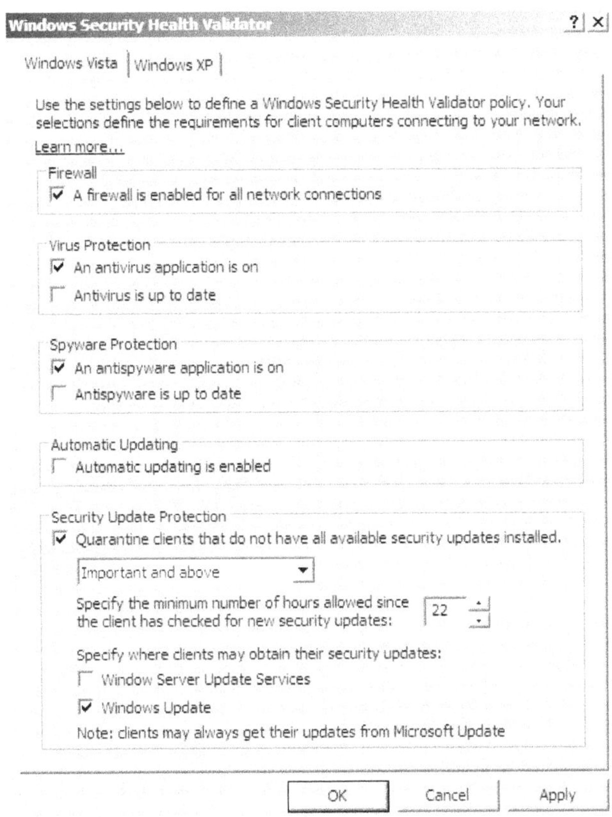

Figure 4.21. *Interface of the Microsoft NAP solution allowing specification of the security policy*

The ESCV looks to have a great future, as evidenced by the growing interest taken by large network equipment manufacturers such as Cisco with the NAC (*Network Admission Control*) initiative, Juniper Networks with UAC (*Unified Access Control*) or software vendors, such as Microsoft with NAP (*Network Access Protection*).

To these different providers of proprietary solutions must be added the TCG consortium (*Trusted Computing Group*) with more than 80 companies as members (McAfee, etc.), which defined the specification for an open and interoperable ESCV solution based on RADIUS and EAP, named TNC (*Trusted Network Connect*).

NOTE.– Even if they are not, strictly speaking, NAC, some SSL/IPSec gateways may also by downloading proprietary software, check the conformity of client (version of patches installed, presence of anti-virus, etc.) before allowing it or not to access corporate resources.

4.5.3. *Mobile NAC*[29]

However, the limitation of ESCV is that it only supports mobile devices when they connect to the organization's LAN, either physically or through a VPN solution.

This is why some companies such as the *Fiberlink Communications Corporation* have extended this model of NAC to maintain the integrity of mobile devices even when they are "in the wild".

The main difference with the traditional NAC solutions lies in the fact that to ensure the autonomy of the system, functions for analyzing compliance with the security policy, possible access restrictions, and updates of the configuration of the operating system and applications, will be performed by an NAC software previously installed on the mobile device and not distributed on one or more servers.

Thus, when the mobile device is connected to the Internet, the embedded NAC software will update its security policy, which in particular, incorporates the latest list of *patches,* which need to be applied. It will

29 The term NAC was coined by Cisco, and entered into common language to describe an ESCV solution. It is in this sense that the term is used in the remainder of this section.

therefore be able to verify a device's compliance, and depending on the result of this study, initiate the appropriate action plan:

– automatic installation of software patches;

– prevention of connections via a "Wi-Fi hotspot" if the antivirus software is not active;

– restricting access to some Internet servers if the OS does not have the latest patch;

– limiting of the Internet connection's authorizations to the email client only, if the personal firewall is not loaded into memory.

However, there is a feature that a mobile NAC cannot offer: the control of access to the LAN for unauthorized users. As you will have understood, in the present state of affairs, mobile NAC solutions do not know how to interact with LAN equipment (switches), which is why both NAC solutions are not mutually exclusive but rather complementary.

NOTE.– With a traditional NAC system the problem of bandwidth does not arise, because the equipment is connected with a speed of at least 10 Mbit/s via LAN. But with a "mobile NAC" solution, things can be quite different. It is therefore important that the "mobile NAC" solution used manages this constraint by allowing:

– a patch download to be resumed from the point where they had previously stopped;

– patches downloaded and installed with priority given to those most critical to security;

– the Quality of Service to be managed so that the connectivity needs of the user (surfing the Internet, receiving emails, etc.) are not disturbed by downloading patches.

Now that we have reviewed, in the context of this book, the main techniques of attack and their associated countermeasures, we can further analyze what Mr. Rowley's IT department should have done in order to avoid all the problems that were mentioned in Chapter 1.

Chapter 5

What Should Have Been Done to Make Sure Mr Rowley's Day Really Was Ordinary

"You can't defend. You can't prevent. The only thing you can do is detect and respond."

Bruce Schneier

(Responsible for the security of BT's technologies)

5.1. The attack at Mr Rowley's house

5.1.1. *Securing Mr Rowley's PC*

You will already have gathered from reading this book that the first line of defense to be implemented is automatic updating of operating systems and software. If Mr Rowley had enabled this feature on his personal computer, the hacker would not have been able to exploit the operating system vulnerability to install the worm

There may still be a lag between the discovery of a vulnerability and the development of the patch to correct it. It is therefore equally important to use antivirus and anti-spyware software that continuously analyzes the computer's memory and the hard disk to detect the presence of malware.

In addition, a personal *firewall* is also recommended, because if despite all these precautions a worm still manages to install itself, the various communication attempts that it makes in order to spread will invariably be blocked by the *firewall*. Furthermore, this will attract attention to the worm and betray its presence.

On a work PC, profiles can be set up to restrict the rights of the average user to prevent them from making changes to operating system settings; most worms need to modify the *registry* (or equivalent on operating systems other than Windows) to be able to install themselves on the target computer, and since making changes to the registry depends on the rights associated with the user, worms will, in most cases, be blocked.

5.1.2. *Securing the organizational level*

As we have seen, the most promising solution for dealing with contamination of an organization's local network by the mobile devices which connect to it, is the NAC. If this had been implemented, Mr Rowley's PC would have automatically been quarantined until such a time as its operating system had been updated and the worm eradicated. Only when his computer was found to comply with his company's security policy would it have been allowed to connect.

Instead of setting up a tunnel solution, it would also have been possible to install a gateway between Mr Rowley's computer and his company's information system (*desktop sharing*, SSL *gateway*, etc.). Because Mr Rowley would have only had access to a virtual environment, it would not have been possible to corrupt the company's information system.

It should be noted that because personal computers are still less reliable in terms of security than those managed directly by an organization's IT department, some companies only allow the latter access to their network in order to avoid such incidents.

If, despite all these precautions, Mr Rowley's PC had still managed to contaminate his company's network, then the IPS would have detected and blocked the worm's propagation attempts.

As a last resort, because some worms use specific TCP/UDP ports, the IT department could have configured the *firewall* that sits between the VPN concentrator and the network, in order to block all packets with this characteristic.

Finally, to limit the spread of a worm, the company's network can be segmented so that in the case of contamination of one part, filtering mechanisms will prevent the infection from spreading to the rest of the network.

5.1.3. *Detection at the organizational level*

The main difficulty for a company faced with infection by a worm is detecting it early. If it is a known worm, or one of its variants, its signature will already be registered in the security equipment databases such as IPS and it will be easily detected. Otherwise, the alarm may be raised only when a large part of the information system has been infected. For this reason, it is preferable to use additional tools such as a *honeypot*.

When the worm attempts to find new hosts in order to "reproduce" itself, it will inevitably attack the *honeypot*, which will have been judiciously placed within the organization's network (at communication crossroads, near the critical production servers, etc.). As nobody accesses the decoy server, it is easy to deduce that this is an attack, and its source can be determined.

An SIEM solution can also be implemented to collect and analyze logs of events that have occurred in the information system in order to find evidence of an intrusion attempt. For example, some worms will try to connect to a machine using a specific range of ports in order to find a vulnerability, or the traffic from an infected computer may suddenly increase; these are a few of the many clues that may indicate the beginning of an invasion.

5.1.4. *A little bit of prevention*

Prevention is also an important element in the battle against data vulnerability. Thus, Mr Rowley and his colleagues should have been made aware of the techniques used by hackers to find victims, strongly discouraged from surfing risky sites (pornography, downloading pirated software, etc.) or using applications which are notorious hubs for such infections (for example, Emule).

5.2. The attack at the airport VIP lounge while on the move

During this training session, the theft of computer equipment and especially the data stored on it should also have been discussed. The golden rule is that you should never leave a computer unattended in a public place, and even on work premises it is recommended that computers are physically attached to the desks during employees' absences so that they cannot be stolen.

However, in our case, the accidental "pirate" did not even take hold of the computer to gain knowledge of its content as Mr Rowley did not take the trouble to close his session before going to the bar in the airport VIP lounge. To avoid such a thing occurring, you should always activate the screensaver so that the computer, after several minutes of inactivity, asks the user to log-in again to get access to it.

If, despite this precaution, a session is compromised, the use of a file encryption solution will maintain data confidentiality, because the hacker would only have had access to encrypted and consequently unreadable data, which will be of no use.

A final precaution would have consisted of the deactivation by an administrator of all the computer's USB ports to avoid data being copied to removable media (USB flash drive, hard disk, or equipment that can behave as a storage device such as an Android smartphone, etc.). Incidentally, disabling the USB ports can also limit contamination by viruses.

To avoid such "information leaks", in addition to these technical solutions, some organizations have classified their documents, and prohibit the files classified as most sensitive from leaving their premises. In our case this would not have been possible, because Mr Rowley needed to have access to confidential information in order to demonstrate the benefits of his product to his customers.

5.3. The attack at the café

When he connected to what he believed to be a Wi-Fi access point, Mr Rowley was in fact snared by a hacker.

Using their DHCP server, the hacker first assigned Mr Rowley an IP address with which to redirect him to the hacker's own DNS server. Thus, regardless of the Web address that Mr Rowley was trying to access, the DNS automatically redirected him to the hacker's relay server. This was achieved by seamlessly relaying Mr Rowley's requests to genuine Web servers. Thus, Mr Rowley was able to connect to an "e-commerce" site to order the new doll for his daughter.

It was only when Mr Rowley prepared to confirm his order that the hacker's relay computer redirected him to a Web server that mimicked the appearance of this "e-commerce" site. This server needed only to ask Mr Rowley to enter his confidential information (password, credit card number, etc.) and to transmit this to the hacker. Once the task was completed, and to avoid arousing Mr Rowley's suspicion, the server sent an error page indicating that following a technical problem the order had not been processed and advising him to try again later. Under certain circumstances, an even simpler attack is possible (not requiring a separate proxy server) where malware on the victim's PC terminates the SSL Connections from the e-commerce site and returns regular non-encrypted HTTP traffic to the local browser. The user must notice the absence of https:// in the browser's address bar, but otherwise the entire transaction will work normally and the order will successfully be submitted.

The best way to determine if a Web server is what it claims to be is to check its certificate. However, few people do this systematically, and for this reason the authentication procedure is integrated into the SSL security protocol.

When Mr Rowley tried to place his order, an SSL session would have been opened, which could have helped draw attention to the deception. However, the hacker's relay computer cleverly intercepted the request and redirected it to a computer that mimicked the appearance of the original server.

Mr Rowley could have discovered this scam if he had been more attentive and had checked that the closed padlock icon, indicating the establishment of an SSL session, had appeared in the bottom right corner of his Web browser.

5.4. The attack in the airport VIP lounge during Mr Rowley's return journey

Finally, when Mr Rowley decided to synchronize his address book between his smartphone and his laptop, he proceeded to link them up for the first time. However, a hacker had set up a PC with Bluetooth listening (*sniffing*) software in the VIP lounge: the hacker knew that such places necessarily attract many people in positions of responsibility, who therefore have easily marketable, confidential information. Through *post hoc* analysis of the exchanges between Mr Rowley's two devices, the hacker could determine their encryption keys and gain access to all the transferred data.

To better understand how the hacker proceeded, it is necessary to explain the way the Bluetooth security model works at the link level. This takes place in three phases:

– *Creation of the initialization key* (Kinit) using the E22 algorithm. To generate this key, the algorithm uses the following parameters: the PIN that was entered by the user on the two devices he wishes to link, the unique address of the slave machine (BD_ADDR_B) and a random number (IN_RAND). This random number is communicated by the master machine to the slave *via* Bluetooth without encryption.

– *Creation of the link key* (Kab) using the E21 algorithm. To create this key the algorithm uses the following parameters: a random number generated by the master machine (LK_RAND_A), a random number generated by the slave machine (LK_RAND_B), the unique address of the master machine (BD_ADDR_A), the unique address of the slave machine (BD_ADDR_B). Both machines first exchange their respective "LK-RAND", encrypting them with the "Kinit" key and the logical operator "XOR" (exclusive OR).

– *Mutual authentication of the two machines* via a challenge-response mechanism. Machine A sends a randomly generated unencrypted message (e.g. AU_RAND_A) to its partner machine B. It then generates a response based on the E1 algorithm that uses the parameters "AU_RANDA", the "link key" (Kab) and its own unique address (for example, BD_ADDR_B). Once it has received machine B's response, machine A, which initiated the exchange, is also able to generate its own "response" and check that the result is identical. Since only a machine which knows the link key can generate this response, this proves its identity. Once the identity of B is proven to A, it is then up to B to challenge A to, in turn, ensures A is what it claims to be.

You will understand after having read this brief explanation of this security mechanism that, by listening passively to exchanges, a hacker can easily obtain:

– the unique address of the master machine (BD_ADDR_A);

– the unique address of the slave machine (BD_ADDR_B);

– the random number (IN_RAND);

– the message generated randomly by the master machine (AU_RAND_A);

– the message generated randomly by the slave machine (AU_RAND_B).

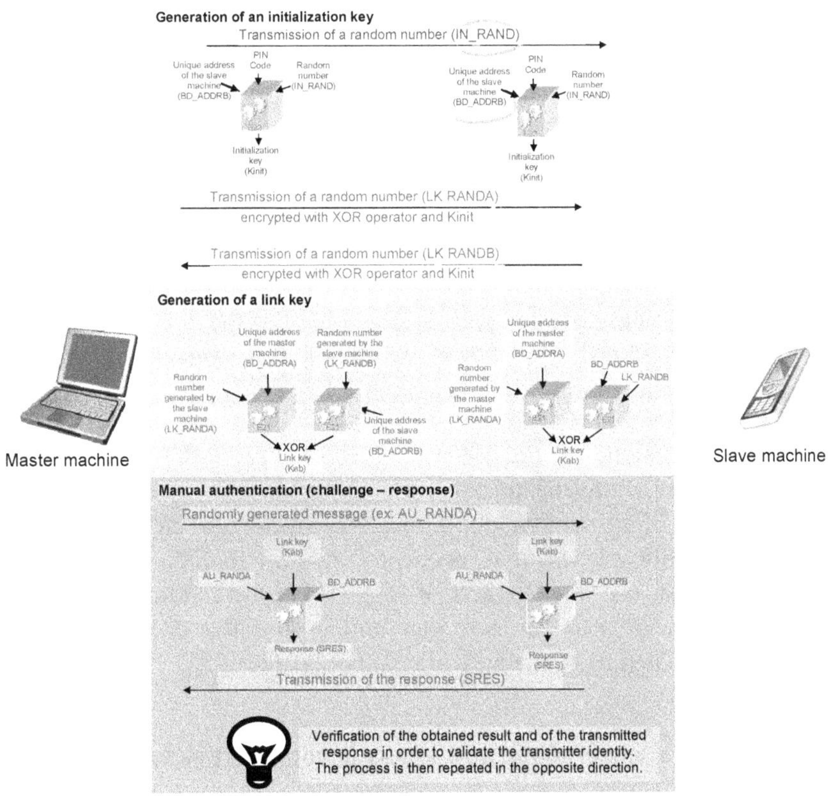

Figure 5.1. *Operation of the Bluetooth security model at the link level*

The hacker then used a *brute force* attack (see section 2.3.4 – *Breaking encrypted data*), generating several potential "Kinits", using all possible combinations of the PIN. These were then used to decode "LK_RAND$_A$" and "LK_RAND$_B$". This then allowed the hacker to generate several potential link keys (Kab). Finally, the hacker was able to determine, from all these potential "Kabs", which was correct, thanks to the mutual authentication process: knowing the messages randomly generated by master and slave machines (AU_RANDA, AU_RANDB), the hacker could test all combinations of "Kab" to see which one obtained the same results as those generated by Mr Rowley's machines.

Mr Rowley's main mistake was to link his devices for the first time in a public place. As we have explained, this allowed the hacker to discover the "IN_RAND" and thus deduce the link key. If Mr Rowley had performed this operation at home or in the office, the two devices would already have disposed of the encryption key and would therefore have been able to protect their communications.

The second error was the choice of PIN. Mr Rowley used four digits, despite the fact that it could include up to sixteen. We have seen that the greater the number of digits in a password, the more combinations are possible, and the more complicated the attacker's task becomes.

The work of Yaniv Shaked and Avishai Wool of the *School of Electrical Engineering Systems* showed that it was possible to find a four-digit PIN code with a 3 GHz Pentium IV in less than 0.1 s.

So in conclusion, if you do not wish, along with numerous other organizations, to compensate for weaknesses in the Bluetooth security protocol by disabling equipment on your employees' devices or prohibiting their use in public places, it is necessary at least to link equipment in a secure area and use the longest PIN possible. In addition specialized encryption software can be used that will ensure that the data captured during passive listening to the network cannot be decoded.

5.5. The loss of a smartphone and access to confidential data

The loss of Mr Rowley's smartphone, and the absence of a PIN code to protect access to it, enabled the data that it contained to be compromised.

The first protection for this type of mobile equipment should be the activation of the device's lock code using a password (numeric or alphanumeric). This security measure offers minimal protection, but is nevertheless sufficient to prevent access due to simple curiosity or by an unskilled thief.

Enabling data encryption capabilities then also minimizes the risks posed by unsolicited access. Finally, all these risks can be avoided by using a remote data wiping solution in the case of loss or theft.

A more comprehensive approach would be to integrate Mr Rowley's smartphone through an MDM solution to automatically deploy the company's adopted security policies to all mobile devices that connect to the information system.

5.6. Summary of the different security solutions that should have been implemented

So that you can get an overview of resources and solutions that would have allowed Mr Rowley to have a truly ordinary day, we conclude this work with a summary of the various attacks that he has suffered, and the countermeasures that should have been implemented by implementing the recommendations of the ISO 27002 standard.

The ISO 27002 standard or "Code of practice for the management of information security" is derived from the work of the *British Standards Institution* on information security (BS 7799-1). This standard identifies and describes the various security measures to be implemented to ensure an information system's security, through the 133 best practice recommendations made by companies in the field.

These recommendations are collected in 11 chapters or sections, the content of which are below:

– section 5: Information security policy;

– section 6: Organization of corporate structure;

– section 7: Asset management;

– section 8: Security associated with Human Resources;

– section 9: Physical and environmental security;

– section 10: Communications and Operations management;

– section 11: Access control;

– section 12: Acquisition, development and maintenance of information systems;

– section 13: Management of security incidents;

– section 14: Management of business continuity;

– section 15: Compliance.

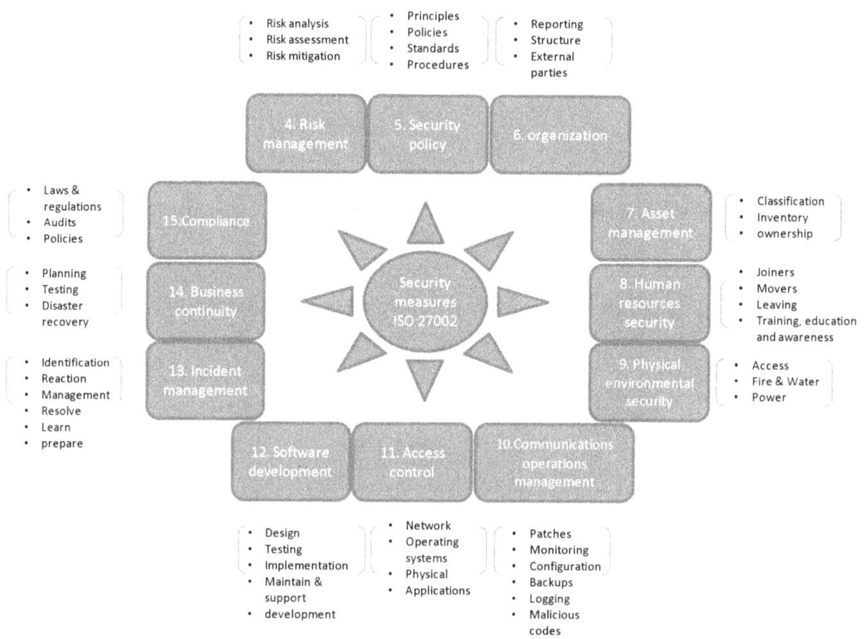

Figure 5.2. *Nomenclature of the ISO/IEC 27002 standard (a) ISO 27001 security home*

Table 5.1 summarizes the different security measures which should have been put in place in the case of Mr Rowley, according to the ISO recommendations.

Attack	Recommended security measure	Corresponding recommendation of ISO/IEC 27002
Contamination by a worm	Automatic installation of patches for the various software installed on the handheld device, including the operating system	**Section 12** (Acquisition, development and maintenance of the information system) paragraph 12.6: Technical vulnerabilities in systems and applications should be checked by examining announcements of security breaches, assessing risk and promptly applying the appropriate patches.
		Section 10 (Communications and operations management) paragraph 10.4: Protection against malicious and mobile code … by anti-*malware* control.
	Use of a *firewall* on the handheld device to prevent external attacks and detect any attempts of unauthorized communication	**Section 10** (Communications and operations management) paragraph 10.6: Recommendations concerning the management of the network, its supervision, and other controls.
	Locking the local account so that the user cannot install software or change the configuration	**Section 11** (Access control) paragraph 11.5: A means of controlling access to the operating system should be used.
	Implementation of a NAC-type system for controlling access to the local network.	**Section 11** (Access control) paragraph 11.4: Access to network services must be controlled from within the organization.
	Use of a gateway (*desktop sharing*, *SSL gateway*, etc.) instead of a VPN solution for accessing the organization's resources.	**Section 11** (Access control) paragraph 11.6 : Access to the application must be checked according to the security policy defined by the company. Particularly sensitive applications may require the use of dedicated isolated platforms.

Attack	Recommended security measure	Corresponding recommendation of ISO/IEC 27002
	Installing of an IPS to automatically detect and eradicate infection attempts.	**Section 13** (Management of security incidents) paragraph 13.1: An alert process is required, as well as procedures associated with incident handling and escalation.
	Implementation of filtering mechanisms/ compartmentalization to limit the spread of infection.	**Section 11** (Access control) paragraph 11.4: Information services, users and systems must be isolated in separate logical networks.
	Use of a *honeypot* to detect infection attempts as early as possible.	**Section 10** (Communications and operations management) paragraph 10.10: It is necessary to record security events and supervise systems in order to detect unauthorized use attempts.
	Implementing an SIEM solution to collect and analyze logs in order to find traces of an intrusion attempt.	**Section 10** (Communications and operations management) paragraph 10.10: It is necessary to record security events and supervise systems in order to detect unauthorized use attempts.
	Education of users about risks posed to information systems, and to the techniques used by hackers.	**Section 8** (Security linked to Human Resources) paragraph 8.2: Employees should be instructed and trained in security procedures.
Theft of information from laptops/ smartphones and tablets	Education of users about risks posed to information systems, and to the techniques used by hackers.	**Section 8** (Security linked to Human Resources) paragraph 8.2: Employees should be instructed and trained in security procedures.
	Automatic locking of the session after a certain period of inactivity.	**Section 11** (Access control) paragraph 11.5: A locking mechanism after a period of inactivity should be applied.

Attack	Recommended security measure	Corresponding recommendation of ISO/IEC 27002
	Use of a file encryption solution to protect the confidentiality of data.	**Section 12** (Acquisition, development and maintenance of information systems) paragraph 12.3: An encryption policy should be defined, specifying roles and responsibilities.
	Disabling of all USB ports on the computer so that data cannot be copied to removable media (USB drive, HDD).	**Section 10** (Communications and operations management) paragraph 10.7: Use of *backup* devices should be recorded and controlled.
	Prohibition of leaving the organization's premises with data categorized as sensitive.	**Section 7** (Asset management) paragraph 7.2: Information must be classified according to protection requirements and labeled accordingly; appropriate control measures should be applied.
Theft of bank information using a "falsified" Web server	Checking the Web server certificate.	**Section 10** (Communications and operations management) paragraph 10.9: The security implications of e-commerce should be assessed and appropriate protective measures implemented.
	Verification of the establishment of an SSL session.	**Section 10** (Communications and operations management) paragraph 10.9: The security implications of e-commerce should be assessed and appropriate protective measures implemented.
Theft of address book	Initialization of linking of Bluetooth devices only in a secure place.	**Section 10** (Communications and operations management) paragraph 10.8: Security procedures and standards must be put in place to protect information in transit.

Attack	Recommended security measure	Corresponding recommendation of ISO/IEC 27002
	Use of a complex password (PIN) which is difficult for a hacker to guess (dictionary attack, *brute force*, etc.).	**Section 11** (Access control) – paragraph 11.3: Users should be made aware of their responsibilities with respect to maintaining the efficacy of access control: choice of complex passwords, etc.
	Deactivating the Bluetooth function in public places.	**Section 10** (Communications and operations management) paragraph 10.8: Security procedures and standards must be put in place to protect information in transit.

Table 5.1. *Available and recommended security measures*

Conclusion

When planning the design and deployment of a remote access solution, following our best practices for safety will dramatically reduce any risk you incur. It is fundamentally important, however, to be aware that there is no such thing as zero risk and it is therefore essential to use multiple means of protection such as the implementation of a multi-level network architecture (mobile workstations, perimeter zone, internal network, application servers).

In addition, once implemented, a security architecture become's obsolete very quickly, as hacker attacks continuously evolve and become increasingly complex. Hackers easily take advantage of new technologies to share information, know-how (forums, specialized Websites, etc.) and to coordinate their actions. Care must be taken to regularly update the infrastructure (applying software patches, installing of new antivirus signatures etc.), frequently carry out audits and to not hesitate, each year, to verify the safety of the global security architecture. This will ensure the network provides updated protection against your identified and important risks, and integrates the latest security requirements and capabilities.

The principal difficulty in the implementation and management of remote access security policies is in maintaining a balance between the constraints imposed by these protection tools and procedures, and the flexibility provided by these mobility solutions. Security officials will tend to minimize connection opportunities to correspondingly reduce risk of intrusion, whereas users, by contrast, wish to have the same level of service and the same ease of access as they have in their office.

Moreover, you should never forget that the first line of defense – which is also, paradoxically, the main "Achilles' heel" of security – is none other than the end user. If he does not adhere to safety procedures, finding them too restrictive, he risks "short circuiting" them and thus jeopardizing the integrity of the information system. Conversely, if he adheres to the security process, he can identify potential attacks and alert the relevant authorities in the case of suspicious behavior (appearance of error messages, SSL padlock icon presence, etc.) and thus enable an attempted attack to be rapidly addressed, even before it is identified by the security mechanisms in place.

Care must therefore be taken to ensure the integrity of the information system, not only by relying on the technical elements (*firewall*, IPS, anti-virus, etc.), but also by involving the user from the very beginning, with training and specifically adapted procedures (acceptable use policy, implementation of a single phone number for any security problems, etc.); because as we have seen, the human being is the keystone of the success of any security policy.

> "If you think technology can solve your security problems, then you don't understand the problems and you don't understand the technology."

Bruce Schneier
Chief Security Technology Officer, BT

Appendices

Appendix 1

Summary of Security Solutions

Risks at the access level	Solutions
Theft of password.	– Use a strong authentication system (OTP, biometric, etc.). – Use a virtual keyboard to avoid keyloggers intercepting the password. – Install anti-malware software on the device to automatically suppress keyloggers. – Encrypt exchanges between remote equipment and the server it wishes to access to avoid sniffing attacks. – Prevent reuse of old passwords. Require users to regularly change their password. – Implement a mutual authentication mechanism between the VPN gateway and the client to avoid spoofing attacks. – Protect hashed password files from unauthorized copying by using shadow links to secured locations.
Identity theft due to brute force or dictionary attack.	– Use a strong authentication system (OTP, biometric, etc.). – Require users to choose a password comprising alphanumeric characters and composing of more than eight numbers or letters. – Prevent reuse of old passwords. Require users to regularly change their password.

	– Implement a delay between unsuccessful authentication attempts that increments at each attempt. – Lock accounts after a certain number of failed authentication attempts within a set amount of time.
Access to resources/information which are not in the user's work domain.	– Implement a policy for accessing the organization's various resources according to user profile (accounts, human resources, etc.). Thus, users can only access the resources necessary for them to carry out their work. – Implement segmentation of the organization's network. – Implement a positive-enrollment system in order to ensure that only authorized persons have access to different resources, and expire that access automatically if it is not used. – Publish an acceptable use policy for computer equipment, clearly stating that employees who do not comply with these rules will face disciplinary action.
Access to the organization's network by a third person from an unattended device.	– Train users on basic security rules. – Enable the screen saver automatically after a certain period of inactivity. – Automatically disconnect the VPN tunnel after a certain period of inactivity.
Access to the organization's network by an individual using a lost or stolen mobile device.	– Use a strong authentication system (OTP, biometric, etc.) which is not simply based on possessing the mobile device. – Implement a procedure for deactivating a remote access account following a simple call from the user.
Access to the organization's network using a replay attack	– Use a VPN solution to avoid the hacker reusing prior connection sessions.
Access to the organization's network using a "man in the middle" attack.	– Implement a mutual authentication mechanism between the VPN gateway and the client.
Theft of authentication information by a hacker, via installation of a "fake" remote access server	– Implement a mutual authentication mechanism between the VPN gateway and the client.

Unavailability of a remote access service.	– Implement an IPS and/or a firewall between Internet access and the equipment responsible for the remote access service (Web portal, dedicated server, etc.) to prevent all attempted saturation attacks. – Do not automatically deactivate an account in the case of multiple unsuccessful authentication attempts, but put in place a delay between two attempted connections. – Implement a load sharing architecture to manage the failure of equipment or sharp rises in requests (saturation attack). – Implement QoS management to prioritize the most important flows. – Deactivate non-essential services which may be used by hackers to saturate the servers (ping).
Risks of information theft at the remote equipment level	
Theft of information contained on mobile device.	– Protect access to the mobile device by a system of identification (biometrics). – Automatically and systematically encrypt all stored data. – Disable the USB ports to prevent a third party from copying the data stored on an unsupervised device. – Enable the screen saver automatically after a certain period of inactivity. – Configure the mobile device so that it cannot save data locally, only on a company server. – Activate the *session CleanUp* function on the VPN client to automatically perform "housekeeping" on the mobile device when the login session is terminated. – Use a virtualization system or remote management system through which the data will be stored on central servers. – Establish a policy for acceptable use of computers prohibiting the release of confidential information outside the company premises. – Train users on basic security rules (never leave a mobile device unattended in a public place). – Install anti-malware software on mobile devices to automatically remove trojans and other spyware.

	– Install a personal firewall to block attempts by spyware to send data. – Prohibit the use of public facilities that may contain spyware unless the VPN solution provides a secure space (virtual office). – Use a data auto-destroy function that can be activated remotely.
Theft of information contained on a removable storage device.	– Automatically and systematically encrypt all data stored on the removable storage medium. – Configure the mobile device (disable USB ports, etc.) so that it cannot save data to a removable storage device. – Train users on basic security rules (always keep the removable storage media such as USB keys with you, etc.). – Disable "AutoRun" type functionality when USB keys are inserted into USB ports.
Reading of information displayed on a screen by a third person by watching "over the shoulder" of the user.	– Use a filter on the screen of the mobile device preventing a person other than the user to read the content that is displayed.
Risks at the transmission level	
Data intercepted or modified during transmission.	– Use an VPN solution to secure exchanges (confidentiality/integrity). – Implement a policy for acceptable computer use prohibiting the use of email to send confidential information (unless it is previously encrypted).
Risks at the level of the violation of integrity of the remote access device	
	– Use a firewall on the mobile device to prevent external attacks. – Install anti-malware on mobile devices to automatically remove viruses/worms, etc. – Automatic installation of various software patches installed on the handheld device, including the operating system. – Automatic updating of signature databases of anti-malware software installed on the handheld device. – Lock local users' accounts so that they cannot install software.

Risks at the level of violation of the integrity of the organization's information system	
Spread to the rest of the information system of malware from an infected handheld device.	– Train users on the basic security rules (do not surf risky Websites). – Install anti-malware on mobile devices to automatically remove viruses/worms, etc. – Systematically and automatically updating operating systems and software installed on the handheld device based on the provision of patches by their vendors. – Systematically and automatically updating signature databases of security tools installed on the mobile device. – Install a firewall on the handheld device to prevent attacks. – Implement an IPS to analyze and process the flows leaving remote connections. – Implement a policy for accessing different resources according to user profiles (accounts, human resources, etc.) so that in the event of contamination, only a limited number of resources are infected. – Automatically check compliance of the mobile device with security rules (anti-virus up to date, no malware detected, etc.) before it is allowed to connect to the corporate network. – Favor the use of solutions based on Web portals which allow access sites via a browser only, if access is made from a device not managed by the company (cybercafe). – Prohibit all connections that are not initiated from a handheld device managed by the company.

Appendix 2

Glossary

0 to 9

802.1x: IEEE protocol controlling access to the LAN network of an organization – both who and how; see *IEEE, LAN, WPA, WPA2.*

A

AAA (*Authentication, Authorization and Accounting*): acronym recalling the three basic functions of servers such as RADIUS systems, which manage authentication and access permissions, as well as logging of connection events for later analysis (recharging for use); see *Kerberos, Radius, TACACS +.*

Access Control List: *see ACL.*

ACL (*Access Control Lists*): filtering rules for network flow, in terms of protocol type, IP addresses and ports that can be implemented on a router or firewall; see *Firewall, Router.*

Active Security: alliance of publishers, manufacturers and consultancies, launched in 1999 by Network Associates, around the Event Orchestror architecture to enable security tools to work together; see *ANSA, OPSEC.*

Active X: technology launched in February 1996 by Microsoft to allow the development of small applications that can be downloaded via the Internet; *see Java, Vandal.*

Address translation: *see NAT.*

Adware: this term is a contraction of *advertising* and *software*, indicating software that delivers advertising messages (pop-ups) and/or collects information that can be used for marketing purposes; *see Spyware.*

AES (*Advanced Encryption Standard*): symmetric key encryption algorithm invented by two Belgian researchers (Joan Daemen of Proton World International and Vincent Rijmen from the University of Louvain), following a tender launched in 1997 by NIST (National Institute of Standards and Technology) to replace DES; *see ANSI, Secret key encryption, DES, RC5, RSA, WPA2.*

AFW (application firewall): *see Firewall application.*

ANSA (*Adaptive Network Security Alliance*): alliance of editors, developers, consultancies and telecoms operators, for the definition of an interoperability norm for security tools relating to the ISS intrusion detection solutions; *see Active Security, OPSEC.*

ANSI (*American National Standards Institute*): American standardization organization founded in 1918 (www.ansi.org). Most notably, it has worked on the C programming language and on the ASCII character system; *see DES.*

Antivirus: software designed to detect and eradicate computer viruses; *see Signature (of a virus), Virus.*

Appliance: "black box" grouping a set of software into a single device to facilitate installation, configuration and ongoing management; *see UTM.*

Application Firewall: program which analyzes the content of messages to applications (Web server, database, etc.) in order to ensure that they do not contain an attack (*Stack overflow*, insertion of SQL code); *see Firewall.*

ARP (*Address Resolution Protocol*): means of knowing the physical address (MAC) of a machine in order to be able to communicate with it using a higher level protocol such as IP; *see MAC address, ARP Poisoning, IP*.

ARP *Poisoning*: method of identity theft involving replacing the MAC address of the attack's "target" machine with that of the hacker in the caches of the computers that will be communicating with it. They therefore unwittingly send all communications to the hacker's machine instead of communicating with the "target"; *see MAC address, IP spoofing.*

Assistant: software which assists the user in completing a specific task, for example configuring the settings of an application.

Asymmetric key encryption: *see public key encryption.*

Authentication certificate: *see Digital certificate.*

B

Back Orifice: program developed in 1998 by the *hacker* group CdC (*Cult of the dead Cow*) to allow controls of Windows 95 and 98 PCs from a remote computer; *see CdC.*

BackDoor: the backdoor is a hidden feature that has been included in a program or an operating system by a developer in order to allow access to certain resources, without having to go through the authentication process; *see RootKit.*

Bandwidth: maximum number of bits that can be transferred per second via a given medium. The units of measurement most commonly used to characterize transmission speed are Kbits/s, Mbits/s and Gbits/s.

Bayesian filter: filter based on a mathematical algorithm, notably used to calculate the conditional probability of an email being spam; *see Spam.*

Biometrics: methods of identity control based on an individual's morphological characteristics.

Black List: list of resources (mail relay, Web pages, etc.) that are prohibited to access and/or from which communications are rejected, because they do not respect an organization's rules (pornographic content, etc.) or which may present some kind of risk (e.g. spam); *see Spam, White list.*

Bluetooth: wireless communication technology (2.4 GHz) invented in 1994 by Ericsson to enable the exchange of information between devices over a short distance.

BotNet: a contraction of the words roBOT and NETwork, which designates a group of "zombie" machines (or Bots) which are remotely controlled by an individual or an organization with a worm or a "Trojan". Once established, these networks allow denial of service attacks to be launched, or spam to be sent; *see Trojan, Droneware, Spam, Worm, Zombie.*

Browser (Web): software for visualization of Web pages and navigation of the Web by following hypertext links; *see HTML, HTTP.*

Brute Force Attack: technique used by hackers to find passwords. It consists of testing all possible combinations until one is found which matches the password used; *see Dictionary attack.*

Bug: error in a program or operating system that can have unforeseen consequences such as causing the machine to "crash", erroneous data processing, security breaches, etc.; *see Patch.*

C

CC (Common Criteria): criteria for evaluating the security features of components of an information system, established in 1993 and standardized by ISO in 1999 (ISO 15408); *see ISO, ITSEC.*

CdC (*Cult of the dead Cow*): the name of a group of hackers known for developing the Back Orifice software. The name was chosen in reference to a speech by Chinese Premier Li Peng during his meeting with Bill Gates; *see Back Orifice.*

Challenge-response authentication: authentication system which uses a server that generates a code (stimulus) that the user must enter into their "calculator". This then calculates a response from the code and confidential information which is shares with the server (shared secret). The user then simply uses the result for authentication; *see Synchronous mode authentication.*

Codebreaking: refers to the act of seeking and finding a password or the encryption key of a message.

Connection Hijacking: hacking technique which consists of waiting until the victim has successfully completed the authentication phase, and then taking their place in the communication.

Cryptography: study of techniques of protecting the content of messages and ensuring their authenticity; *see Steganography.*

Cybercrime: criminal activity perpetrated in Cyberspace.

Cyberspace: word invented by the science fiction writer William Gibson in his novel *Neuromancer* (1984), which described a society based on a computer network. By extension, we also use this term to designate the Internet or all other virtual worlds; *see Internet.*

D

Datagram: in order to carry data over the network, the communication protocol (e.g. TCP/IP) groups them into virtual envelopes (called datagrams or packets) to which the information necessary for their transit are added, such as the destination address; *see IP, TCP.*

DDoS (*Distributed Denial of Service*): attack method which consists of using many computers to "bombard" a target machine with requests (email, ping, etc.) in order to saturate it and thus prevent it from providing its intended service; *see BotNet, Denial of Service, Zombie.*

Defacer: hacker who changes the content of Websites; *see Hacker.*

Demilitarized zone: *see DMZ.*

Denial of service: technique used by hackers that consists of sending thousands of requests (usually TCP or ICMP) to a server which, overwhelmed by this traffic, can no longer provide its service, or can only do so with reduced quality (response time, etc.); *see DDoS, ICMP, Land, Lock flood, Ping of death, Scanning, Smurfing, TCP, WinNuke.*

DES (*Data Encryption Standard*): symmetric key encryption algorithm invented in 1974 by IBM, then standardized in 1981 by ANSI; *see AES, ANSI, Secret key encryption, RC5, RSA.*

DHCP (*Dynamic Host Configuration Protocol*): IETF protocol allowing dynamic assignment of an IP address to a machine; *see IP Address, IETF.*

Dictionary Attack: technique used by hackers to find passwords. It consists of using a database of commonly-used words (first name, celebrity names, etc.) and testing them all, until one is found that matches the password used; *see Brute force attack.*

Digital certificate: electronic document containing the necessary information (cardholder name, public key, name of the certification body, etc.) to authenticate the identity of its owner (application, server, user); *see Public key encryption, PKI, Escrow.*

Digital ID: *see Digital certificate.*

Digital identity card: *see Digital certificate.*

Distributed Denial of Service: *see DDoS.*

DMZ (*Demilitarized Zone*): buffer zone between the outside (usually the Internet) and an organization's network, controlled by a firewall. Here, only company servers to communicate with the outside and the reverse proxies are installed. The rest of the information system is therefore not exposed to potential attacks by hackers, and only the DMZ machines can potentially be compromised; *see Firewall, Proxy, Reverse Proxy.*

DNS (*Domain Name System*): system invented in 1983 (RFC 882, 883, 1987, 1034, 1035) ensuring association between a computer and its IP address; *see DNS poisoning, DNS spoofing.*

DNS poisoning: technique for corrupting the buffer memory of certain versions of DNS servers (BIND v4, BIND v8, etc.) by substituting the IP addresses that are stored in those computers with those of computers controlled by the hacker; *see DNS, DNS spoofing.*

DNS spoofing: method of identity theft which consists of replacing, in a DNS server, the IP address of a machine with that of another, which is controlled by the hacker; *see DNS, DNS poisoning.*

Droneware: type of *spyware* specialized in creating Botnets (network of computers controlled from a distance by hackers); *see Botnet, Spyware.*

Dynamic password: *see One Time Password.*

E

EAP (*Extensible Authentication Protocol*): extension of the PPP protocol which allows several methods of user authentication to be offered; *see PPP.*

EFF (*Electronic Frontier Foundation*): American association, founded in July 1990, which aims to defend fundamental civic liberties in cyberspace.

Electronic courier: *see Email.*

Email: this term is a contraction of the words *electronic* (e) and *mail* (mail), designating an asynchronous message that can be sent from a computer; *see SMTP, Spam, worm.*

ESCROW: organization which guarantees the identity of the correspondents in a communication relationship, using digital certificates, *see Digital certificate.*

ESM (Enterprise Security Manager): *see SIM.*

F

False alarm: alarm generated by a surveillance system when the event it is supposed to have detected has not actually occurred.

False positive: *see False alarm.*

Firewall: device placed between the private network of the company and external connections (Internet, partners, etc.) to prevent intrusion attempts. Only flows that have been specified in the configuration of this device are allowed to pass; *see DMZ, application firewall, NAT, OPSEC, UTM.*

Freeware: program of which use and copying is free of copyright; *see Shareware.*

FTP (*File Transfer Protocol*): protocol for transferring files, based on TCP (port 20 and 21); *see TCP.*

H

Hacker: individual who succeeds in fraudulently entering an application or an operating system to use resources (CPU time, disk space, etc.), view or modify sensitive information (credit card numbers, bank account, etc.) or affect proper functioning; *see Defacer.*

Hashing algorithm: algorithm for calculating the signature of a file to ensure it has not been modified since its creation.

HIDS (*Host-based Intrusion Detection System*): software installed on servers and clients to continuously analyze the activity of the operating system and automatically generate an alarm (SNMP trap, email, SMS, etc.) in case of detection of anomalous behavior; *see HIPS.*

HIPS (*Host-based Intrusion Prevention System*): software installed on servers and clients to deal with attack attempts; *see HIDS.*

HIS (Head of Information Security): person who is responsible for the security of the information systems at the core of an organization.

Hot Spot: Internet access *via* Wi-Fi which is offered in a public place (café, train station, etc.).

HTML (*Hyper Text Markup Language*): description language of Web pages and of hypertext links, derived from SGML, which was conceived in 1989 by Tim Berners-Lee; *see HTTP, SGML.*

HTTP (*Hyper Text Transfer Protocol*): file transfer protocol in unconnected mode, used notably for downloading HTML pages; *see HTML, HTTPS, URL.*

HTTPS (*Secure* HTTP): extension of the HTTP protocol, allowing the addition of certain security functions (signature, encryption, authentication); *see HTTP, SSL.*

I

ICMP (*Internet Control Message Protocol*): protocol (IETF RFC 792) which provides a set of features to assist in the management of IP networks, notably used by the ping and traceroute utilities; *see Denial of Service, IETF, IP, Ping, Ping of Death, Ping Sweep, TraceRoute.*

ICMP scan: *see Ping sweep.*

IDMEF (*Intrusion Detection Message Exchange Format*): standardized message format based on XML, which has been defined by the IETF to allow the interaction of various security devices, in the case of detection of an attempted attack; *see IDS, IPS.*

IDS (*Intrusion Detection System*): security system that continuously analyzes traffic and automatically generates an alarm (SNMP trap, email, SMS, etc.) if it detects abnormal behavior; *see IDMEF, IPS, SNMP, SMS, UTM.*

IEEE (*Institute of Electrical and Electronics Engineers*): standardization organization in the telecommunications, electronics, electrical and computing domains, which was founded in 1963, but with origins dating back to the late 19th Century; *see RPR, Wi-Fi, WIMAX.*

IETF (*Internet Engineering Task Force*): standardization organization for technologies based on IP, which was founded in 1986; *see DHCP, ICMP, IDMEF, IKE, IP, IPSEC, L2TP, LDAP, RFC, TCP, TLS.*

IKE (*Internet Key Exchange*): mechanism (RFC 2409) used by IPSEC to allow partners in a communication to exchange encryption keys; *see IPSEC.*

Instant message: system of communication which allows real-time exchange of message in text format; *see SPIM, Worm.*

Internet: global communication infrastructure, based on the IP protocol, of which the foundations were laid in 1969 (ARPANET) by the Pentagon, who at that time wanted to build a network capable of resisting a nuclear attack; *see Cyberspace, IP.*

IP (*Internet Protocol*): packet-type network protocol which is most notably used with UDP and TCP; *see Datagram, ICMP, IETF, Internet, IP spoofing, PPP, TCP, UDP, VOIP.*

IP Address: 4-bit encoded (four numbers from 0 to 255) address used by version 4 of the IP protocol to identify all devices connected to the network; *see DHCP, IP, RFC 1918.*

IPSEC: IETF standard designed to bring together identification, authentication and encryption functions, missing until the instigation of the IP protocol; *see IKE, IP, L2F, VPN.*

IP Spoofing: method of identity theft consisting of using the IP address of a trusted machine; *see ARP Poisoning, IP.*

IPS (*Intrusion Prevention System*): security system that continuously monitors traffic to detect any abnormal behavior, in order to automatically implement previously-specified defense procedures (reconfiguration of the firewall, cutting the communication session, etc.); *see IDMEF, IDS, Firewall, UTM.*

ISO (*International Organization for Standardization*): organization created in 1947 to federate the national standards organizations of more than 140 countries; *see CC, JPEG, OSI Model, SGML.*

ISO 8879: *see SGML.*

ISO 15408: *see CC.*

ISP *(Internet Service Provider)*: company allowing businesses and individuals to connect to the Internet.

ITSEC (*Information Technology Security Evaluation Criteria*): criteria for assessing the adequacy and effectiveness of the security features of information system components, originally developed in 1980 by France, UK, Germany and Holland; *see CC.*

ITU (International Telecommunications Union): international agency, responsible for the standardization of technologies linked to telecommunications, which brings together 184 member states; *see H323.*

J

Java: interpretive, object-oriented programming language, which is inspired by C++ and *smalltalk*. It was created in 1995 by James Gosling for SUN, which required a programming tool that was independent of operating systems (Mac OS, Windows, Unix, etc.); *see Active X, Vandal.*

Java Applet: small applications written in JAVA downloaded by a client to be executed by a virtual Java machine. They are commonly used to include animations on Web pages or to extend the functionality of Web browsers; *see Active X, Vandal.*

Jitter: transit time variation during the transmission of multiple packets belonging to the same communication session.

JPEG (*Joint Photographic Expert Group*): compressed image format invented by ISO in 1990 which is based on the fact that image quality can be degraded, because the human eye cannot distinguish all nuances of color and detail; *see ISO.*

K

Kerberos: authentication system developed by MIT under the framework of the Athena project; *see MIT, RADIUS, TACACS+.*

Keylogger: this term is a contraction of *keystroke* and *logger*, which means recording, unbeknownst to the user, the sequences of characters typed into the keyboard, in order to intercept sensitive information such as passwords, credit cards number, etc. This is a type of Spyware; *see RootKit, Spyware.*

Kiddiot: pejorative version of the term Script kiddy; *see Script kiddy.*

L

L2F (*Layer Two Forwarding*): Cisco proprietary protocol, allowing creation of level 2 tunnels; *see IPSec, L2TP, PPTP, VPN.*

L2TP (*Layer Two Tunneling Protocol*): IETF standard produced by the fusion of the proprietary protocols PPTP and L2F, allowing the creation of level 2 tunnels (RFC 2661, RFC 3931); *see IPSec, L2F, PPTP, VPN.*

LAN (*Local Area Network*): local network which is deployed over a limited area (generally a building or a campus); *see 802.1x, VLAN, WAN.*

Land: software which appeared in 1997 that allowed a packet with the same IP address and port number in the source and destination fields to be sent to a machine. The computer that received this packet therefore perpetually re-sent it to itself, causing saturation of its resources which could ultimately block the machine; *see Ping of death, WinNuke.*

Layer 1 of the OSI model: part of the OSI model, known as the physical layer, responsible for the transmission of data on physical media (fiber, copper, etc.); *see Layer 2 of the OSI model, Layer 3 of the OSI model, Layer 4 of the OSI model, OSI model.*

Layer 2 of the OSI model: part of the OSI model, known as the data link layer, responsible for ensuring that the data transmitted by the communication medium arrive safely at their destination (error detection and correction, frame, etc.); *see Layer 1 of the OSI model, Layer 3 of the OSI model, Layer 4 of the OSI model, OSI model.*

Layer 3 of the OSI model: part of the OSI model, known as the network layer, responsible for managing addresses and data routing; *see Layer 1 of the OSI model Layer 2 of the OSI model, Layer 4 of the OSI model, OSI model.*

Layer 4 of the OSI model: part of the OSI model, known as the transport layer, responsible for the management of end-to-end communications (correcting errors, session management, packet retransmission in case of loss or corruption, etc.); *see Layer 1 of the OSI model, Layer 2 of the OSI model, Layer 3 of the OSI model, OSI Model.*

LDAP (*Lightweight Directory Access Protocol*): a protocol for accessing and managing data contained in a directory, developed in 1993 by the University of Ann Arbor (Michigan) from ISO X500 and standardized in 1995 by the IETF (RFC 1777); *see IETF.*

Linux: UNIX operating system which was developed, originally, in 1991, by Linus Torwarlds, who then made it available to the *open source* community, so that everybody could participate in its evolution.

Lock flood: denial of service technique that consists of generating numerous entries in logfiles in order to saturate disk space, therefore causing the system to "crash"; *see Denial of service, Log.*

Log: file or database in which events are automatically recorded (saturation disk space, access attempts, etc.) relating to the operation of a software or operating system. These data can then be used later for auditing, debugging, chargeback, and so on; *see Lock flood.*

Logic bomb: "Trojan" that only becomes active after a particular event (a given date, appearance of a name in a list, etc.) which then causes maximum damage (formatting hard disk, file deletion, data corruption, etc.) in systems to which it has access; *see Trojan, Spyware.*

M

MAC Address (*Media Access Control*): address contained in each network card, identifying each machine; *see ARP, ARPPoisoning.*

Macro language: programming language included in office tools (word processing, spreadsheet) to produce small programs (macros) to automate certain actions; *see Macro virus.*

Macro virus: type of computer virus, written in macro language, which usually spreads by infecting email attachments; *see Macro language, Virus.*

Malware: refers to the collection of software that has been designed to have a harmful effect, such as viruses, *spyware*, Trojans, worms, etc.; *see logic bomb, Trojan, Script kiddy, Spyware, Worm, Virus.*

MIME (*Multipurpose Internet Mail Extension*): IETF standard for sending any type of data (binary files, etc.) in emails and not be limited to 7-bit ASCII characters; *see SMIME.*

MIT (*Massachusetts Institute of Technology*): American university located in Cambridge, Massachusetts, USA, which specializes in research in the scientific and technical domains; *see Kerberos, RSA.*

MODEM (MOdulator/DEModulator): device for the conversion of bits into waves, and the inverse, allowing transmission of data using analog links.

MPPE (*Microsoft Point to Point Encryption*): encryption algorithm developed by Microsoft, based on RC4, which is most notably used to secure data exchange in the PPTP protocol; *see PPTP, RC4.*

MSSP (*Managed Security Service Provider*): company offering their customers an infrastructure security management service, thus allowing businesses without sufficient resources (in number and/or in qualification) to ensure 24/7 security, or to redeploy their teams to tasks with greater added value.

N

NAT (*Network Address Translation*): mechanism generally implemented at the level of routers or *firewalls* to convert one IP address into another. This technique allows hackers to hide the organization's true address scheme, but also allows computers that do not have official IP addresses or are not compatible with RFC 1918 to access the Internet; *see Firewall, IP, RFC 1918 Router.*

NewsGroup: virtual forum dedicated to a particular theme (security, TV series, recipes, etc.) where users can share their written opinions, information, advice, etc.

Neural network: computer system which aims to imitate the structure of the nervous system in order to reproduce the learning, reflection and decision-making of a human being.

O

Open source: software of which the source code is publicly available, which ensures that it conforms to recommendations and does not contain any hidden functions.

Operating system: software which manages the different resources of a computer (hard disk, memory, etc.) and provides an interface (text or graphical) so that the user can access it.

OPSEC (*Open Platform for Secure Enterprise Connectivity*): alliance of publishers, manufacturers, and service providers aiming to define a standard for interoperability of security tools around the *CheckPointFirewall*; *see Active Security, ANSA, Firewall.*

OSI Model (*Open Systems Interconnect*): model developed by ISO to allow communication between different types of computers. It is based on decomposition into seven layers to manage the various communication functions; *see Layer 1 of the OSI model, Layer 2 of the OSI model, Layer 3 of the OSI model, Layer 4 of the OSI model, ISO.*

OTP: *see One time password.*

One time password: system allowing generation of a "disposable" password, which cannot be used more than once, for accessing computing resources; *see Password.*

P

P2P (*Peer-to-Peer*): popularized by software such as Napster, Gnutella, etc., which allow users to exchange files (including audio), this type of architecture is characterized by the fact that each machine is both client and server at the same time (that is, there is no hierarchy between computers); *see Worm.*

PABX (*Private Automatic Branch eXchange*): private branch exchange that connects the various telephones within an organization to one another, and with the outside. Users have access to many features such as voicemail, conference calls, referral orders, etc.; *see RTC TOIP.*

Password: one of the older means of authentication, which consists of sharing secret information with the person or the system in charge of identity verification; *see Unique usage password.*

Patch: correction made to a program or an operating system; *see Bug.*

PDA (*Personal Digital Assistant*): electronic devices which includes a small set of applications (calendar, address book, etc.) allowing their owners to always have the information necessary for managing their organization to hand. It is generally believed that the first PDA was the Apple Newton, released in 1993, which at the time was not as successful as hoped; *see Smartphone.*

Phishing: term derived from the word *fishing*, which appeared for the first time in 1996 on the *hacker* site alt2600. It refers to a technique used by hackers which involves appearing to the victim as a trusted organization (bank, insurance, etc.) to make it easier to elicit confidential information (credit card details, password); *see DNS poisoning.*

Ping (*Packet Internet Groper*): utility from the world of UNIX for testing a connection and calculating the time taken for a packet to be routed between two machines; *see ICMP, Ping of death, Traceroute.*

Ping of death: denial of service technique, consisting of sending an ICMP packet whose size exceeds 65,535 bytes (the limit imposed by RFC 791) which can cause a TCP stack overflow and, more generally, an operating system "crash"; *see Denial of Service, Land, WinNuke.*

Ping sweep: refers to the act of sending ICMP messages to all IP addresses in a defined subnet in order to obtain a list of machines that are active; *see ICMP, TCP scan.*

PKI (*Public Key Infrastructure*): infrastructure for handling requests, allocation, storage and revocation of public keys; *see Public key encryption, digital certificate.*

Port: number associated with the TCP and UDP protocols, which allows unique identification of a service (Telnet, FTP, NFS, etc.) offered by a computer; *see Scanning.*

PPTP (*Point to Point Tunneling Protocol*): proprietary protocol of 3Com, Ascend, Microsoft and US Robotics, allowing creation of level 2 tunnels (RFC2637); *see IPSec, L2F, L2TP, MPPE, VPN.*

PPP (*Point to Point Protocol*): IETF protocol (RFC 1661) for packet transfer via serial links, created in 1994; *see EAP, IP.*

Proxy: device placed between a company's private network and external connections (Internet, partner, etc.) to relay requests from applications. This screens the internal structure of the network from the outside world and, like all terminals that communicate across it, uses a single IP address; *see DMZ Reverse Proxy.*

Proxy server: see *Proxy.*

Promiscuous: particular mode in which the network card of a computer is able to intercept and read all network packets, even those for which the machine is not the destination.

Public key encryption: encryption system which uses two keys. The first, known to all, allows messages to be encrypted, and the second, which is possessed only by the addressee, is used to decipher it. This system uses more complex algorithms than symmetric key encryption, and thus requires much more processing time, but unlike the latter it does not require prior exchange of keys to communicate; *see symmetric key encryption, PKI, RSA.*

R

RADIUS (*Remote Authentication Dial-In User Service*): system for authentication and authorization management, as well as connection event logging, based on a client-server architecture type and on the UDP protocol. It was developed by Livingston (which was subsequently acquired by Lucent) and adopted in 1997 by the IETF (RFC 2865); *see AAA, Kerberos, TACACS +, UDP.*

RC4 (Rivest's Cipher 4): secret key encryption algorithm developed in 1987 by M. Rivest; *see Secret key encryption, DES, MPPE, RC5, RSA, TKIP, WEP*.

RC5 (Rivest's Cipher 5): secret key encryption algorithm developed in 1995 by M. Rivest; *see Secret key encryption, DES, RC4, RSA*.

Reflection Attack: *see Smurfing*.

Registry: database used by Windows operating systems to store information such as license number, passwords, etc.

Reverse proxy: device, generally found in the DMZ, which relays requests coming from the Internet to an organization's internal Web server, to prevent all attack attempts directed from the exterior; *see DMZ, Proxy*.

RFC (*Request For Comments*): document used in the IETF standardization procedure to specify a new technology which could be used on the Internet; *see IETF, RFC 1918*.

RFC 793: *see TCP*.

RFC 821: *see SMTP*.

RFC 822: *see SMTP*.

RFC 882: *see DNS*.

RFC 883: *see DNS*.

RFC 1034: *see DNS*.

RFC 1035: *see DNS*.

RFC 1510: *see Kerberos*.

RFC 1661: *see PPP*.

RFC 1777: *see LDAP*.

RFC 1918: IETF specifications concerning the classes of IP address which can be used in the heart of an organization (private address); *see IETF, NAT, RFC.*

RFC 1987: *see DNS.*

RFC 2246: *see TLS.*

RFC 2409: *see IKE.*

RFC 2865: *see RADIUS.*

Rlogin: UNIX service which allows remote connection to a machine.

Rollback: functionality allowing a system to be returned to its preceding state in the case of malfunction.

Root Kit: software for obtaining, unlawfully, any or all of the admistrator privileges on a machine. Once these privileges are obtained, it will generally conceal itself so as not to be detected, thus enabling the user to discreetly commit misdemeanors (installation of a backdoor, a *keylogger*, etc.); *see Backdoor, Keylogger.*

Router: device whose function is to route data flow through a network so that it reaches the correct recipient; *see ACL, NAT.*

RSA: public key encryption algorithm whose name is drawn from the initials of the three MIT researchers (Rivest, Shamir and Adelman) who participated in its development in 1977. It is based on the mathematical principle that is easy to multiply two integers together (result forming the public key), but there are no easy ways to decompose a large number into a product of prime factors (forming the secret key); *see Public key encryption, DES, MIT, RC5.*

S

SAML (*Security Assertion Markup Language*): protocol based on XML, created by OASIS in order to allow different systems of authentication to exchange data, with the aim of federating identity management.

Sandbox: secure environment that allows applications to be tested to ensure that it will not attempt to perform hostile actions.

SATAN (*Security Analysis Tool for Auditing Networks*): historically, the first automatic information security auditing software. It was conceived by Dan Farmer in 1995.

Saturation Attack: *see Denial of Service.*

Scanning: technique used by hackers that consists of checking whether or not each UDP and TCP port is in use; *see Denial of service, Port.*

Script kiddy: teenager using programs and other scripts available on the Internet to attempt to penetrate information systems, or generate new *malware*; *see Malware.*

Secret key encryption: encryption system which uses the same key to encrypt and decrypt data. This system is faster than asymmetric key encryption, but has the notable disadvantage of requiring a secure communication channel so that the transmitter and the receiver can exchange the key; *see AES, Public key encryption, DES, RC4, RC5.*

Session theft: *see Connection Hijacking.*

SGML (*Standard Generalized Markup Language*): language dedicated to the management of documents in electronic format which was standardized by ISO (ISO 8879); *see HTML, ISO.*

Shareware: software, for the use of which the author requests some kind of involvement from the user (money, a gift, etc.), generally at a reduced rate; *see Freeware.*

Signature (of a virus): sequence of bits allowing identification (as with DNA) a given virus; *see Antivirus, Virus.*

Site defacing: refers to the modification by a hacker of the content of a Website

Smartphone: PDA-type device possessing communication functions via the cellular network; *see PDA.*

SMIME (*Secure* MIME): protocol developed originally by RSA Data Security, allowing authentication and encryption functionalities to be added to MIME.

SMS (*Small Message System*): small message in text format which can be exchanged by mobile telephones; *see IDS.*

SMTP (*Simple Mail Transfer Protocol*): protocol (RFC 821, 822) developed in 1982 for managing the sending of electronic mail; *see email.*

Smurfing: DOS technique which consists of replacing the IP address of a machine and using this false identity to send requests to a multitude of computers, which then provokes a denial of service by responding to this massive demand; *see DOS.*

Sniffing: technique allowing theft of confidential information that consists of capturing packets transmitted via the network for later extraction of their content.

SNMP (*Simple Network Management Protocol*): network management protocol based on UDP, allowing information on the device (number of bits emitted, operating system version, etc.) to be obtained, and warning messages (Trap) to be received; *see IDS, Trap, UDP.*

Social engineering: technique used by hackers to obtain information and sometimes to cause their victims to perform certain actions, by pretending to be someone else (security service, system administrator, etc.).

Spam: refers to the sending of bulk email advertising to the mailboxes of people who have not previously expressed a desire to receive such messages; *see Black List, Bayesian, SPIM, White List, Zombie.*

SPIM (*Spam over Instant Messaging*): sending of unsolicited advertizing *via* instant messages; *see Instant Messaging, Spam.*

Spoofing: technique which consists of usurping the MAC or IP address of a given machine.

Spyware: term referring to the collection of software that is installed on a computer to capture information (credit card number, password, personal files, etc.) unbeknownst to the user. It therefore includes *keyloggers* and *adware*; *see Adware, Keyloggers*.

SSID (*Service Set Identifier*): identifier which serves to name a wireless network.

SSL (*Secure Socket Layer*): secure connection protocol developed by Netscape in 1995, and which is now in the public domain. Because it sits between the application layer and the transport layer (TCP, UDP), it is completely independent of applications (HTTP, Telnet, etc.). It allows, in addition to its ability to encrypt information being transmitted via the network, server authentication as well as that of the client (optional); *see HTTPS, TLS*.

SSO (*Single Sign On*): software ensuring the interface between the different authentication systems so that the end user can access all resources, only having to enter a single password.

Steganography: term meaning "recovered writing" in Greek, which refers to the technique consisting of dissimulating information by hiding it in a vehicle such as an image or audio file; *see Cryptography*.

Strong authentication: authentication not based on a single test mode but a combination of several to exploit the specific advantages of each. Authentication with password generator is an example of strong authentication, since it requires a physical device (which I have) and a code (that I know) to be able to use it.

Switched: LAN environment based on the use of switches so that machines can communicate directly with one another without sharing the transport medium, as was the case with earlier versions of Ethernet; *see LAN, VLAN*.

Symmetric key encryption: *see Secret key encryption*.

Synchronous authentication mode: authentication system which uses a server and a "calculator" owned by the user, which automatically generates, at regular intervals (usually 30 seconds), a code based on the date, time and a shared secret. The user simply has to enter the code displayed on the LCD to authenticate.

T

TACACS+ (*Terminal Access Controller Access Control System*): system for authentication as well as authorization management and accounting (recording logon events), based on a client-server architecture type and TCP. It was originally developed by BBN Planet Corp. for the U.S. Department of Defense, but today it is Cisco that ensures it continues to evolve; *see Kerberos, RADIUS, TCP.*

TCP (*Transmission Control Protocol*): transport protocol based on IP defined by RFC 793; *see Denial of service, FTP, IETF, IP, RFC, TACACS +, UDP.*

TCP scan: discovery technique used by hackers, consisting of sending requests to a specified range of TCP ports, in order to obtain a list of those that are available; *see Ping sweep.*

Time synchronized authentication: *see Synchronous mode authentication.*

TKIP (*Temporal Key Integrity Protocol*): protocol offering encryption of data exchanges on Wi-Fi networks through the use of the RC4 algorithm, like WEP. Unlike the latter, however, it allows the use of temporary keys (128 bits), renewed every ten thousand packages; *see RC4, Wi-Fi, WPA.*

TLS (*Transport Layer Security*): name given to the SSL protocol, standardized by IETF (RFC 2246). Version 1.0 of TLS corresponds to version 3.1 of SSL; *see IETF, SSL.*

Trace Route: utility from the world of Unix revealing through which IP devices (routers) packets have passed while reaching their destination; *see ICMP, Ping, Unix.*

Trap: error message emitted by a device to a supervisory platform using the SNMP protocol; *see SNMP.*

Trojan program: *see Trojan.*

Trojan: program inspired by the ruse used by the Greeks to invade Troy. Specifically, under the guise of a harmless activity (e.g. a game) this software secretly performs a number of illicit activities, such as the transmission of confidential files over the Internet; *see BotNet, Dialer, Zombie*.

Tunnel: *see VPN*.

U

UDP (*User Datagram Protocol*): transport protocol based on IP which, in contrast to TCP, does not control transmissions; *see IP, RADIUS, SNMP,TCP*.

Unix: multi-task, multi-user operating system invented in 1969 by Ken Thompson at Bell laboratories; *see Linux,TraceRoute*.

Unsolicited mail: *see Spam*.

URL (*Uniform Resource Locator*): address format used on the Web for uniquely specifying the route of a resource (an image, an HTML page, etc.).

UTM (*Unified Threat Management*): *appliance* grouping the majority of security software (*firewall*, anti-virus, anti-spam, IDS/IPS, Web URL filtering, etc.) which are necessary to an organization, thus facilitating their installation, configuration and day-to-day management; *see Appliance, Firewall, IDS, IPS, URL*.

V

Vandal: program (Java, ActiveX), developed with the aim of damaging the data contained in the computer that executes it; *see ActiveX, Java, Virus, Worm*.

Virus: program which infects other software, inserting a copy of itself into the program; *see Antivirus, Macro virus, Signature (of a virus), Vandal, Worm*.

VLAN (*Virtual LAN*): local area network realized by configuring multiple Ethernet switches to allow machines that are connected to communicate as if they were connected to a single hub, irrespective of distance; *see switched LAN switched*.

VPN (*Virtual Private Network*): virtual private network generally existing under a public infrastructure such as the Internet, using an encryption solution which ensures confidentiality in data exchanges.

W

WAN (*Wide Area Network*): network to connect multiple LANs separated by large geographical distances (region, country); *see LAN*.

Web Spoofing: technique described for the first time in an article published in December 1996 by Princeton University, which consists of applying the *man in the middle* technique to Web servers.

WEP (*Wired Equivalent Privacy*): security protocol forming part of the Wi-Fi IEEE 802.11 network standard, which unfortunately is not free from design flaws. These principally concern the way in which the RC4 encryption algorithm and the associated key or initialization vector are used. The latter is easy to decrypt, following a passive capture of several million packets; *see RC4, Wi-Fi, WPA, WPA2.*

White List: list of computer resources (message relays, Web pages) which respect an organization's rules and with which it is therefore possible to exchange; *see Black list.*

Wi-Fi (*Wireless Fidelity*): local, wireless Ethernet technology, standardized by IEEE (802.11a, 802.11b, 802.11g, 802.11n). See *SSID, TKIP, WEP, WPA, WPA2.*

Worm: program which spreads from computer to computer by repeatedly reproducing (duplicating) itself, and using vehicles as varied as email, instant messaging, P2P networks, etc.; *see BotNet, Email, Vandal, Virus, Zombie.*

WPA (*Wi-Fi Protected Access*): intermediate solution designed pending the development of the WPA2 standard to address the weaknesses of WEP. It is based on the use of the TKIP protocol for encryption and data integrity control as well as implementing of 802.1x and EAP for authentication; *see 802,1x, EAP, TKIP, WEP, Wi-Fi, WPA2.*

WPA2 (*Wi-Fi Protected Access v2*): integral part of the Wi-Fi 802.11i standard which comprises protocols 802.1x and EAP for mutual authentication between the client and the AAA server, as well as AES for exchange security; *see 802,1x, AAA, AES, EAP, WEP, Wi-Fi, WPA.*

Z

Zombie: computer contaminated by a Trojan or a worm, of which a hacker can take control in order to use it to commit misdeeds (DDoS, sending spam, etc.); *see BotNet, DDoS, Spam, Trojan, Worm.*

Bibliography

[ALT 06] ALTARIS, Remote access best practices: confronting the potential security threats of teleworks, 13 June 2006.

[APP] APPGATE, Remote access with AppGate

[APPa] APPLE, iPhone OS – Enterprise Deployment Guide – 2nd Edition, for Version 3.2 or later, www.manuals.info.apple.com

[APPb] APPLE, Mobile Device Management, Deploying iPhone and iPad, www.images.apple.com.

[ARM 11] ARMSTRONG T., MASLENNIKOV D., "Kaspersky – Android Malware is on the rise", *Virus Bulletin Conference*, Barcelona, Spain, 6 October 2011.

[ATT 01] AT&T, IP Remote Access Dial Security, 17 October 2001.

[BAK 10] BAKKER M., VAN DER JAGT R., GPU-based password cracking, 5 February 2010.

[BEN 03] BENVENUTO M.C., KEROMYTIS A.D., Easy VPN: IPsec Remote Access Made Easy, Computer Science Department, Columbia University, USA, 2003.

[BLA 06] BLACK B., Secure remote access – Technical solution guide, Nortel, January 2006.

[BLU 08] BLUE COAT, Hidden dangers in the mobile worker jungle, 2008.

[BTA] BT ASSURE, Threat Monitoring, The quick and easy way to a healthy network, June 2012.

[CAL 04] CALÉ S., Sécurité informatique : virus, risques et parades, Biotop, 2004.

[CAR] CARNUT M.A., GONDIM J.J.C., "ARP spoofing detection on switched ethernet networks: a feasibility study", *Proceedings of the 5th Simposio Seguranca em Informatica*, November 2003.

[CIS 06] CISCO, Securing your business with your network: security made simple, 2006.

[CIS 06] CISCO NETWORKERS, Understanding, Preventing, and Defending Against Layer 2 Attacks, 2006.

[CIS 08] CISCO, Remote access VPNs: business productivity, deployment and security considerations, February 2008.

[CIS 09] CISCO, Cisco secure remote access solution, 2009.

[CIT 07] CITRIX, Citrix GotoMyPc corporate technology – A simpler approach to secure remote access, 2007.

[CLA 77] BRECHT CLAERHOUT M., A Short Overview of IP Spoofing, 1996-1977.

[COU 05] COUPEL A., "Comment le cryptage quantique veille sur les codes secrets", *ZDNet France*, 3 February 2005.

[DAT] DATAMONITOR, Enabling Enterprise Mobility with Network Access Control, August 2006.

[DOM 05] DOMAGE E., Infrastructures d'accès, une solution architecturale aux nouveaux besoins des entreprises, IDC, October 2005.

[DOM 06] DOMAGE E., "Etat des lieux du marché français de la sécurité des systèmes d'information des entreprises", *Les assises de la sécurité et des systèmes d'information*, 2006.

[ECK 10] Eckersley P., Burns J., "An observatory for the SSLiverse", *Defcon 18*, Las Vegas, USA, July 2010.

[F5] F5 NETWORKS INC., Enterprise remote access, September 2005.

[FEA 01] FEARNOW M., NORTHCUTT S., FREDERICK K., COOPER M., *Intrusion Signatures and Analysis*, Sams Publishing, January 2001.

[FER 05] FERNANDEZ-TORO A., *Management de la sécurité de l'information. Implémentation ISO 27001*, Editions Eyrolles, 2005.

[FEW 07] FEWER S., Security weaknesses inherent in the design of TCP over IP, Harmony Security, 2007.

[FIL 09] FILLIOL E., *Les virus informatiques : théorie, pratique et applications*, Collection IRIS, 2009.

[FRA 05a] FRANCHIN F., MONNET R., *Le business de la cybercriminalité*, Hermès, 2005.

[FRA 05b] FRANKEL S., KENT K., LEWKOWSKI R., OREBAUGH A.D., RITCHEY R.W., SHAMA S.R., Guide to IPsec VPNs, NIST, December 2005.

[FRA 08] FRANKEL S., HOFFMAN P., OREBAUGH A.D., PARK R., Guide to SSL VPNs, NIST, July 2008.

[FSE] F-SECURE, Mobile Threat Report, Q4 2011.

[GRA 12] GRACE M., ZHOU W., JIANG X., SADEGHI A.-R., Unsafe Exposure Analysis of Mobile In-App, Department of Computer Science, Center for Advanced Security Research, North Carolina State University Technical University Darmstadt, March 2012.

[HAR 04] HARLÉ T., SKRABACZ F., *Clés pour la sécurité des SI*, Hermès Science Publications, 2004.

[HAR 08] HARDIKAR A., BAMBENEK J.C.A., Malware 101 – Viruses, SANS Institute, April 2008.

[HER 03] HERZBERG A., DIMACS Security & Cryptography - Crash Course – day 4, Internet Cryptography Tools, Part I: TLS/SSL, Computer Science Department, Bar Ilan University, 2003.

[HOF 08] HOFFMAN D.V., *Implementing NAP and NAC Security Technologies*, Wiley Publishing Inc, 2008.

[JAN 08] JANSEN W., SCARFONE K., Guidelines on cell phone and PDA security, NIST, October 2008.

[JIA 12] JIANG J., LIANG J., DUAN H.N., WU J., LI K., LI J., Ghost Domain Names: Revoked Yet Still Resolvable, Network Research Center, Tsinghua University, MDEA Networks, www.isc.org/files/imce/ghostdomain_camera.pdf, February 2012.

[LAS] LASSERRE X., KLEIN T., Réseaux Privés Virtuels – VPN, www.frameip.com/vpn.

[LED 09] LEDUC G., "Securing TCP conections", Computer security a top down approach, www.montefiore.ulg.ac.be/~leduc/cours/ISIR/GSRI-ch4.pdf, April 2009.

[LEV 12] LEVILLAIN O., SSL/TLS : état des lieux et recommandations, ANSSI, 2012.

[LIN 03] LINLAUD D., Sécurité de l'information : Elaboration et gestion de la politique de l'entreprise suivant l'ISO 17799, Association Française de Normalisation (AFNOR), September 2003.

[LOG] LOGAN A., Rethinking remote access: pervasive enterprise mobility using remote access points, Aruba networks, white paper, 2007.

[MCA] MCAFEE, Enforcing endpoint policies for network access, February 2006.

[MCA 06] MCAFEE, Securing your endpoints for network access with McAfee policy enforcer, February 2006.

[MES] MESSAGELABS, Web use & remote workers: managing the risk, 2009.

[MIC 03] MICROSOFT, Windows server 2003 remote access overview, March 2003.

[MIC 04] MICROSOFT, Secure enhancements for remote access at Microsoft, March 2004. [NEO 07] NEOACCEL, NeoAccel SSL VPN-Plus – Le futur des réseaux privés virtuels, 2007.

[MIC 08] MICROSOFT, Secure Web and remote access enablement, 2008.

[MIT 03] MITNICK K., The Art of Deception: Controlling the Human Element of Security, Wiley, October 2003.

[NET 10] Comment fonctionnent les logiciels permettant d'espionner les mobiles?, 01net magazine, 8 March 2010.

[NZE 04] NZEKA G., La protection des sites informatiques face au hacking, Hermès, 2004.

[OHI 09] Ohio supercomputer center, Remote access security, 2009.

[PHI] PHILLIPS M., Remote access guidance, Office of eHealth Standards and Services, http://logrhythm.com/portals/0/pdf/HIPAA_SecurityGuidanceforRemoteUseFinal.pdf.

[PHI 09] PHIFER L., "SSL VPN grows up: time to demand more from your next SSL VPN", *WatchGuard*, May 2009.

[POU 03] POUGET F., DEBAR H., Honeypot, honeynet, honeytoken: terminological issues, Eurocom, white paper, 14 September 2003.

[SAN] Sans institute, Remote access policy, 2006.

[SAN 08] SANTOS O., *End-to-End Network Security Defense-in-Depth*, Cisco Press, 2008.

[SCA 08] SCANSAFE, Roaming workers – The weakest link in corporate web security, April 2008.

[SCA 09] SCARFONE K., HOFFMAN P., SOUPPAYA M., Guide to enterprise telework and remote access security, NIST, June 2009.

[SCH] SCHNEIER B., Cryptanalyse des extensions d'authentification PPTP de Microsoft (MS-CHAPv2)

[SCH 01] SCHNIER B., *Cryptographie appliquée, Algorithmes, protocoles et codes source en C*, Vuibert, 2001.

[SCH 08] SCHUDEL G., SMITH D.J., Router Security Strategies, Cisco Press, Indianapolis, USA, 2008.

[SHA 05] SHAKED Y., WOOL A., Cracking the Bluetooth PIN), www.eng.tau.ac.il/~yash/shaked-wool-mobisys05/, 2005.

[SON 08] SONICWALL, Clean VPN approach to secure remote access, 2008.

[SPA 04] Spafford E.H., The Internet Worm Program: An Analysis, Department of Computer Sciences Purdue University, 2004.

[SPI 02] SPITZNER L., Honeypots - Definitions and Value of Honeypots, www.enteract.com/~lspitz, 17 May 2002.

[SPI 03] SPITZNER L., Honeypots: Catching the Insider Threat, August 2003.

[STA] STAR, Remote access: Three steps to getting connected, white paper, March 2008.

[STE 94] STEVENS W.R., *TCP/IP Illustrated - Volume 1, The Protocols*, Addison-Wesley Professional, 1994.

[SUN 02] SUN Microsystems, Secure remote access with the Solaris 9 operating environment, 2002.

[THY 10] THYER J., Spoofing ICMP redirect host messages with hping, research paper, 26 May 2010.

[TOM 06] TOMUR E., DEREGOZU R., GENE T., A wireless secure remote access architecture implementing role based access control: WiSeR, World academy of science, Engineering and technology, 2006.

[TRA 04] TRABELSI Z., *La sécurité sur Internet*, Hermès, 2004.

[UNI] University of Pittsburgh, How to meet the health check secure remote access service security requirements, report, http://technology.pitt.edu/Documents/network/secure-remote-access/SSL_VPN_Requirements.pdf.

[VLA 05] VLADIMIROV A.A., GAVRILENKO K.V., MIKHAILOVSKY A.A., WI-FOO – *Piratage et défense des réseaux sans fil*, Campus Press, www.wi-foo.com, June 2005.

[WAG 97] WAGNER D., Schneier B., Analysis of the SSL 3.0 protocol, University of California, Berkeley Counterpane Systems, 15 April 1997.

[WAT] WATCHGUARD, Implementing an identity and access management strategy for the mobile enterprise, June 2008.

[WEI 06] WEITH L., Differences Between SSLv2, SSLv3, and TLS, analysis report, 3 July 2006.

[ZEM 00] ZEMOR G., *Cours de cryptographie*, Cassini, 2000.

Further Reading

Journals

General journals

01 Informatique, www.01net.com/01informatique/

01 Réseaux, www.01net.com/01reseauxdirect

Internet Professionnel

Le monde informatique, www.weblmi.com

NetSurf

Planète Internet

Réseaux & Télécoms, www.reseaux-telecoms.net

ZDNet, www.zdnet.fr

Journals on security

Confidentiel Securité, www.confidentiel-securite.com

CSO, www.csofrance.com

Mag-secures, www.mag-securs.com

Netcost & Security, www.netcost-security.com

SC magazine, www.scmagazine.com/home/index.cfm

Security Magazine

MISC

Journals on hacking

Hackademy Magazine

Hackers Magazine

Hacking

Pirates Mag

Universal Hacker

Web servers

Web servers on security

ActuSecu, www.actusecu.info

Biometrie, biometrie.online.fr

BlackHat, www.blackhat.com

CCCure.net, www.cccure.net

Cisco design guides, www.cisco.com/en/US/netsol/ns742/networking_solutions_program_category_home.html

Cit@delle, citadelle.intrinsec.com

Clusif, www.clusif.fr

Common Vulnerabilities and Exposures, www.cve.mitre.org/cve/

Cyberworld Awareness Security Enhancement Structure, www.cases.public.lu

Encyclopédie Symantec sur les virus, securityresponse.symantec.com/avcenter/vinfodb.html

(in)Secure Magazine, www.insecuremag.com

ISO 27001 security home, www.iso27001security.com

National Vulnerability Database, nvd.nist.gov

Netstumbler, http://www.netstumbler.com/

Remote Exploit, www.remote-exploit.org/index.php/Main_Page

Secuser.com, www.secuser.com/index.htm

Securité.org, www.securite.org/index2.html

Sécurité Info, www.securiteinfo.com

Security Focus, www.securityfocus.com

Serveur thématique sur la sécurité des systèmes d'information, www.ssi.gouv.fr/fr/index.html

The center for Internet security, www.cisecurity.org

Virus Bulletin, www.virusbtn.com/index

Virtual Private Network Consortium, www.vpnc.org

WildList, www.wildlist.org

Zataz, www.zataz.com

Zone-H, www.zone-h.fr

Hackers websites

2600[1], www.2600.com

Chaos Computer Club, www.ccc.de/?language=fr

Websites specializing in other types of threat

Infoguerre, http://www.infoguerre.com

Other websites

CCM (IT encyclopedia), www.commentcamarche.net

Comment ça marche, www.commentcamarche.net/attaques/attaques.php3

IANA (list of TCP and UDP ports), www.iana.org/assignments/port-numbers

Wikipedia, www.wikipedia.org

Newsletters

Bugtraq

Full Disclosure

News Groups

alt.2600

alt.security

comp.security.announce

comp.security.firewalls

comp.security.misc

comp.virus

---

1 This name comes from the frequency (2,600 Hz) which was used by the maintenance services of telephone operators in the USA in order to avoid charging for calls. It was hijacked by *phreakers* so they could telephone for free.

Other sources

CNIL, Guide du correspondant informatique et libertés

CNIL, Guide pratique les employeurs, October 2005.

CNIL, Rapport sur la cybersurveillance sur les lieux de travail du 5 février 2002, March 2004.

CSI/FBI Computer Crime and Security Survey, Computer Security Institute, 2006.

La sécurité des systèmes d'information. Un enjeu majeur pour la France, Député P. Lasbordes, November 2005.

MEDEF, Guide de sensibilisation à la sécurisation du système d'information, mai 2005.www.medef.fr/staging/site/core.php?pag_id=36442

Index

M, N, O

P

R

S

T, V

W, Z

CPSIA information can be obtained
at www.ICGtesting.com
Printed in the USA
JSHW042257240621
16220JS00002B/3

9 781848 214354